T0182103

Reduced-Order Modeling (ROM) for Simulation and Optimization

Winfried Keiper · Anja Milde
Stefan Volkwein
Editors

Reduced-Order Modeling (ROM) for Simulation and Optimization

Powerful Algorithms as Key Enablers
for Scientific Computing

 Springer

Editors
Winfried Keiper
Department of Corporate Research
Robert Bosch GmbH
Renningen
Germany

Stefan Volkwein
Fachbereich Mathematik
Universität Konstanz
Konstanz
Germany

Anja Milde
Interdisciplinary Center
  for Scientific Computing
Heidelberg University
Heidelberg
Germany

ISBN 978-3-030-09199-6   ISBN 978-3-319-75319-5   (eBook)
https://doi.org/10.1007/978-3-319-75319-5

Printed on acid-free paper

This Springer imprint is published by the registered company Springer International Publishing AG
part of Springer Nature
The registered company address is: Gewerbestrasse 11, 6330 Cham, Switzerland

Preface

KoMSO Workshop

The growing demand for numerical solutions in scientific computing (modeling, simulation, data analysis, and optimization problems in many application fields) requires ever-higher algorithmic and computational performance.

Why is this so? To name just a few reasons: Models become larger and require higher geometrical resolution and full 3D topology. Larger, more comprehensive systems with challenging boundary conditions are being modeled and ask for robust process control. Inverse problems of much larger size need to be solved. Connected components with more relevant physical effects are simulated simultaneously. Optimization with parametric variants is performed. Problems require multi-domain and multi-scale modeling. An analysis of very large "big data" sets is required. Highly complex and much finer optimization criteria are applied… All these trends will persist for the foreseeable future.

There is a solution to the growing needs for scientific computing power on the one hand: the progress in High-Performance Computing (HPC) hardware. We do not know how much longer Moore's Law will be valid, before microelectronics reaches a physical limit. The other lever for improved computational performance, with immense opportunities, is the algorithmic side of scientific computing.

It has been shown in many cases that the advancement of mathematical algorithms has increased the algorithmic performance clearly more than the improvement of computer hardware alone. In the past, approved examples of massive reductions in computational efforts by fast algorithms include the fast Fourier transform, or the solution of Poisson's equation, or more recently the application of adjoint variable methods. In many cases, "Math can do more than Moore."

As a side effect, new, faster simulation and optimization tools that can run on existing computer hardware will enable many additional applications, especially for problems that cannot be solved numerically today at affordable cost and effort. This will attract many more future users: a contribution to the digital transformation of R&D.

With this motivation in mind, this workshop was organized. It gathered creators and developers of new, optimal, and fast mathematical algorithms and the industrial users of these tools.

The workshop was supported by the Forschungsverbund Wissenschaftliches Rechnen, Baden-Württemberg (WiR-BaWü), a network of various departments in the state of Baden-Württemberg ranging from mathematics, to engineering to computer science. The state of Baden-Württemberg played a pioneering role in advancing scientific computing. It started with the Interdisciplinary Center for Scientific Computing (IWR) pursuing its vision to develop mathematical and computational methods for applications in engineering, the sciences, and humanities.

This proceeding issue summarizes the successful meeting, which took place in November 2016, at the premises of Robert Bosch Corporate Research in Renningen near Stuttgart. These are the major groups of contributions to the workshop:

Recent advances in Model Order Reduction (MOR) simulations: In many applications (e.g., in computational fluid dynamics), the spatial discretization yields quadratic-bilinear parameterized descriptor systems. *Benner and Goyal* extend recent advances in interpolation-based optimal control MOR, ensuring bounded errors for their approximation. Vectorial kernel approximations are very important, because they arise in MOR approximations for systems with vectorial inputs and outputs. *Haasdonk and Santin* prove quasi-optimal rates of convergence for vectorial kernel greedy algorithms.

New MOR based optimization approaches: In many practical applications, many objectives are equally important and also contradictory, such that one is forced to find an optimal compromise between them. This results in multi-objective optimal control problems where multiple objectives have to be minimized simultaneously. This again requires many numerically expensive target evaluations. To overcome the computational burden, *Beermann et al.* develop a MOR strategy, which is utilized in a derivative-free set-oriented optimization solver. A-posteriori error bounds ensure the desired accuracy of the ROM. *Heinkenschloss and Jando* present a method to solve linear quadratic optimal control problems by adjusting a suitable ROM for the Hessian in each level of the iterative method. The computed optimal controls have the same accuracy as the ones obtained by the high-fidelity approximation while being computationally much faster.

Successful bridges to the industrial world: In the context of energy optimization, the simulation of gas networks has become more and more important recently. However, many nonlinear Euler equations have to be solved on the edges of the network. Due to the high complexity of the network, MOR offers the chance to decrease the degrees of freedom dramatically. *Benner et al.* first approximate the Euler equations by a differential–algebraic system of index 1, which can then be successfully reduced by MOR. The simulation of electric rotating machines is both computationally expensive and memory-intensive. For that reason *Bontinck et al.* develop a MOR strategy that handles the challenging non-symmetric setting in the problem. The fast numerical simulation of elastic multi-body systems is of great interest for industrial applications. *Fehr et al.* introduce the program package

Morembs, available in C++ and in MATLAB. The software is already used in various applications. *Hartmann et al.* discuss MOR for digital twins. They point out that MOR is a key technology to transfer sophisticated simulation models into other domains and life cycle phases.

The workshop was jointly organized by KoMSO, Konstanz University and the Robert Bosch GmbH and co-sponsored by the Federal Ministry of Education and Research (BMBF).

Renningen, Germany   Winfried Keiper
Konstanz, Germany   Stefan Volkwein

Contents

An Iterative Model Reduction Scheme for Quadratic-Bilinear Descriptor Systems with an Application to Navier–Stokes Equations

Peter Benner and Pawan Goyal

Abstract We discuss an interpolatory model reduction framework for quadratic-bilinear (QB) descriptor systems, arising especially from the semi-discretization of the Navier–Stokes equations. Several recent results indicate that directly applying interpolatory model reduction frameworks, developed for systems of ordinary differential equations, to descriptor systems, may lead to an unbounded error between the original and reduced-order systems, e.g., in the $\mathscr{H}_2$-norm, due to an inappropriate treatment of the polynomial part of the original system. Thus, the main goal of this article is to extend the recently studied interpolation-based optimal model reduction framework for QB ordinary differential equations (QBODEs) to aforementioned descriptor systems while ensuring bounded error. For this, we first aim at transforming the descriptor system into an equivalent ODE system by means of projectors for which standard model reduction techniques can be applied. Subsequently, we discuss how to construct optimal reduced systems corresponding to an equivalent ODE, without requiring explicit computation of the expensive projection used in the analysis. The efficiency of the proposed algorithm is illustrated by means of a numerical example, obtained via semi-discretization of the Navier–Stokes equations.

1 Introduction

High-fidelity modeling of dynamical systems is often required to have a better understanding of the underlying dynamical behaviors of a system. However, numerical simulations of such high-fidelity systems are expensive and often inefficient. Thus, it is not a straightforward task, or sometimes not even possible, to perform engineering and control design studies using these high-fidelity systems. One approach to circumvent this problem is *model order reduction* (MOR), aiming at constructing

P. Benner · P. Goyal (✉)
Max Planck Institute for Dynamics of Complex Technical Systems,
Sandtorstraße 1, 39106 Magdeburg, Germany
e-mail: goyalp@mpi-magdeburg.mpg.de

P. Benner
e-mail: benner@mpi-magdeburg.mpg.de

© Springer International Publishing AG, part of Springer Nature 2018
W. Keiper et al. (eds.), *Reduced-Order Modeling (ROM) for Simulation and Optimization*,
https://doi.org/10.1007/978-3-319-75319-5_1

1

surrogate models (reduced-order models) which are less complex and replicate the important dynamics of the high-fidelity system.

In this paper, we investigate MOR for quadratic-bilinear (QB) descriptor systems of the form

$$E_{11}\dot{v}(t) = A_{11}v(t) + A_{12}p(t) + Hv(t) \otimes v(t) + \sum_{k=1}^{m} N_k v(t) u_k(t) + B_1 u(t), \qquad \text{(1a)}$$

$$0 = A_{21}v(t) + B_2 u(t), \quad v(0) = v_0, \qquad \text{(1b)}$$

$$y(t) = C_1 v(t) + C_2 p(t), \qquad \text{(1c)}$$

where $E_{11}, A_{11}, N_k \in \mathbb{R}^{n_v \times n_v}, k \in \{1, \ldots, m\}, H \in \mathbb{R}^{n_v \times n_v^2}, A_{12}, A_{21}^T \in \mathbb{R}^{n_v \times n_p}, B_1 \in \mathbb{R}^{n_v \times m}, B_2 \in \mathbb{R}^{n_p \times m}, C_1 \in \mathbb{R}^{q \times n_v}, C_2 \in \mathbb{R}^{q \times n_p}; v(t) \in \mathbb{R}^{n_v}$ and $p(t) \in \mathbb{R}^{n_p}$ are the state vectors; $u(t) \in \mathbb{R}^m$ and $y(t) \in \mathbb{R}^q$ are the control input and measured output vectors of the system, respectively, and $u_k(t)$ is the k-th component of the input vector; $v_0 \in \mathbb{R}^{n_v}$ is an initial value for $v(t)$. Furthermore, we assume that E_{11} and $A_{21}E_{11}^{-1}A_{12}$ are invertible. Hence, the linear part of the system (1) ($H = 0, N_k = 0$) has an index-2 structure, e.g., see [20]. The structure of the QB descriptor system (1) occurs after the space discretization of a control problem where the state functions are described by the incompressible Navier–Stokes equations.

MOR techniques for linear systems are now very well-established and are widely applied in numerous applications, e.g., see [2, 4, 14, 26]. Several of those techniques have been successfully extended to special classes of nonlinear ODE systems, namely bilinear and QB systems, see, e.g., [6, 7, 9, 10, 12, 16, 18, 27]. These techniques can be classified mainly into two categories: trajectory-based methods and system-theoretic approaches. The primary ideas of trajectory-based methods rely on a set of snapshots of the state solutions for training inputs, which is then used to determine a *Galerkin* projection, for more details, see, e.g., [3, 15, 22, 24, 25]. On the other hand, in system-theoretic approaches, there are two prominent methods, the so-called balanced truncated and moment-matching (interpolation) methods that are widely used, see, e.g., [2]. The idea of balanced truncation is to find states which are hard to control as well as hard to observe, and truncating such states gives us a reduced-order system. This method for QB systems has been recently studied in [12]. Interpolation-based methods aim at constructing reduced-order systems which approximate the input–output behavior of the system. With this intent, such a problem for QB systems was first considered in [18], where a one-sided projection method to obtain an interpolating reduced-order system is proposed. Later on, a similar problem was addressed in [8, 9] for single-input single-output (SISO) QB systems, where a two-sided projection method was proposed, ensuring a higher number of moment to be matched. However, the main challenges for this method are a good selection of interpolation points and the application to multi-input multi-output (MIMO) QB systems. To address these issues, an interpolation-based optimal model reduction problem for QB systems was recently addressed in [13], where a reduced-order system is constructed, aiming at minimizing a system norm of the error system, where a truncated $\mathcal{H}_2$-norm is suggested for this purpose.

However, little attention has been paid to descriptor systems (DAEs) which involve algebraic constraints as well as a differential part, and this still requires further research. For descriptor systems, there is a need of structured approaches, since a straightforward application of interpolation techniques to descriptor systems may lead to an unbounded error in a system norm, e.g., the $\mathcal{H}_2$-norm, see, e.g., [19] for linear systems. In the direction of MOR for nonlinear DAEs, interpolation-based methods for specially structured bilinear DAEs have been investigated, e.g., in [11, 17]. Furthermore, a moment-matching method for SISO QB systems as in (1) ($q = m = 1$) was studied in [1]. However, the first challenge in this method is the choice of interpolation points which plays a crucial role in determining the quality of a reduced-order system, and secondly, it is applicable only to SISO systems which are certainly very restrictive from a real-world application point of view.

In this work, we aim to extend the $\mathcal{H}_2$-optimal model reduction framework for QBODEs [13] to QBDAEs, having the structure as in (1). To that end, we first recall an interpolation-based $\mathcal{H}_2$-optimal model reduction technique for QBODEs and the corresponding iterative scheme to construct reduced-order systems, in the subsequent section. In Sect. 3, we then present a transformation of the system (1) into an equivalent ODE system by means of projectors so that the iterative scheme can be applied. We further investigate how the iterative scheme can efficiently be applied to the equivalent ODE systems to obtain reduced-order systems without computing the projectors explicitly. Finally, in Sect. 4, we illustrate the proposed methodology using a lid-driven cavity model, which is obtained by semi-discretized Navier–Stokes equations.

2 Model Reduction for Quadratic-Bilinear ODEs

In this section, we briefly discuss an $\mathcal{H}_2$-optimal model reduction problem for QBODEs. We begin by introducing the problem setting for QBODEs. For this, we consider a QB system given in state-space form:

$$\Sigma : \begin{cases} \dot{x}(t) = Ax(t) + Hx(t) \otimes x(t) + \sum_{k=1}^{m} N_k x(t) u_k(t) + Bu(t), & \text{(2a)} \\ y(t) = Cx(t), \quad x(0) = 0, & \text{(2b)} \end{cases}$$

where $x(t) \in \mathbb{R}^n$, $u(t) \in \mathbb{R}^m$, and $y(t) \in \mathbb{R}^l$ are the state, input, and output vectors at the time instant t; all other matrices are real and are of appropriate dimensions. It is assumed that the matrix A is stable. Furthermore, one may also consider a general nonsingular matrix E in front of $\dot{x}(t)$; however, to keep the discussion simple in this section, we consider it to be an identity matrix.

In the context of model reduction, our aim is to replace the system (2) with a simpler and reliable reduced system, having the form:

$$
\hat{\Sigma} : \begin{cases} \dot{\hat{x}}(t) = \hat{A}\hat{x}(t) + \hat{H}\hat{x}(t) \otimes \hat{x}(t) + \sum_{k=1}^{m} \hat{N}_k \hat{x}(t) u_k(t) + \hat{B}u(t), & \text{(3a)} \\[2mm] \hat{y}(t) = \hat{C}\hat{x}(t), \quad \hat{x}(0) = 0, & \text{(3b)} \end{cases}
$$

where $\hat{x}(t) \in \mathbb{R}^r$ with $r \ll n$, all other reduced matrices are of appropriate dimensions, and the reduced output $\hat{y}(t)$ approximates the corresponding original output $y(t)$ for all admissible L_2-bounded system inputs.

We focus on constructing the reduced system (3) by means of Petrov–Galerkin projection. For this, we need to identify two appropriate projection matrices $V \in \mathbb{R}^{n \times r}$ and $W \in \mathbb{R}^{n \times r}$ such that the state $x(t)$ can be approximated as $x(t) \approx V\hat{x}(t)$, where $\hat{x}(t)$ is the reduced state and satisfies a Petrov-Galerkin condition as follows:

$$
W^T \left(V\dot{\hat{x}}(t) - AV\hat{x}(t) + HV\hat{x}(t) \otimes V\hat{x}(t) + \sum_{k=1}^{m} \hat{N}_k \hat{x}(t) u_k(t) + Bu(t) \right) = 0.
$$

Assuming $W^T V$ is invertible, this yields a reduced system (3) with the realization computed as

$$
\hat{A} = (W^T V)^{-1} W^T A V, \qquad \hat{N}_k = (W^T V)^{-1} W^T N_k V, \quad k \in \{1, \ldots, m\},
$$
$$
\hat{H} = (W^T V)^{-1} W^T H (V \otimes V), \qquad \hat{B} = (W^T V)^{-1} W^T B, \qquad \hat{C} = CV.
$$

As noted in the introduction, there are several methods proposed in the literature to determine these projection matrices V and W or the corresponding subspaces. However, our primary objective is to extend the $\mathscr{H}_2$-optimal model reduction technique [13] to the QBDAE system (1). Before proceeding further, we first recall important properties of the vec $(\cdot)$ operator, stacking the columns of a matrix on top of each other:

$$
\text{vec}\,(XYZ) = (Z^T \otimes X)\,\text{vec}\,(Y), \tag{4}
$$

where $\otimes$ denotes the Kronecker (tensor) product. Secondly, we note down an important concept from tensor theory, that is, matricization of a tensor, see, e.g., [23].

Definition 1 Consider a K-th-order tensor $\mathscr{L} \in \mathbb{R}^{I_1 \times I_2 \times \cdots \times I_K}$. Then, the *mode-k matricization* of the tensor $\mathscr{L}$, denoted by $\mathscr{L}^{(k)}$, is determined by mapping the elements $(i_1, i_2, \ldots, i_K)$ of the tensor onto the matrix entries (i_k, j) as follows:

$$
j = 1 + \sum_{m=1, m \neq k}^{K} (i_m - 1) J_m, \quad \text{where} \quad J_m := \prod_{g=1, g \neq m}^{m-1} I_g.
$$

Using the above concept, we can construct a third-order tensor $\mathcal{H}^{n \times n \times n}$ such that its mode-1 matricization is the same as the Hessian H in (2). Also, let $\mathcal{H}^{(2)}$ and $\mathcal{H}^{(3)}$ denote the mode-2 and mode-3 matricization of the tensor $\mathcal{H}$, respectively. Furthermore, without loss of generality, we can assume that $H(w_1 \otimes w_2) = H(w_2 \otimes w_1)$ for given vectors $w_1, w_2 \in \mathbb{R}^n$, see [9].

Having said that, we note down the definition of the $\mathcal{H}_2$-norm and its truncated version, the so-called truncated $\mathcal{H}_2$-norm for QB systems (2), see [13].

Definition 2 Consider a QB system (2). Assuming the $\mathcal{H}_2$-norm of the system exists, then the $\mathcal{H}_2$-norm is defined as

$$\|\Sigma\|_{\mathcal{H}_2} := \sqrt{\operatorname{tr}\left(\sum_{i=1}^{\infty} \int_0^{\infty} \cdots \int_0^{\infty} f_i(t_1, \ldots, t_i) f_i(t_1, \ldots, t_i)^T \right)},$$

in which

$$f_i(t_1, \ldots, t_i) = C g_i(t_1, \ldots, t_i), \tag{5}$$

where the $g_i(t_1, \ldots, t_i)$ satisfy

$$g_1(t_1) = e^{At_1} B,$$
$$g_2(t_1, t_2) = e^{At_2} \left[N_1, \ldots, N_m \right] \left(I_m \otimes g_1(t_1) \right),$$
$$g_i(t_1, \ldots, t_i) = e^{At_i} \Big[H \left[g_1(t_1) \otimes g_{i-2}(t_2, \ldots, t_{i-1}), \ldots, g_{i-2}(t_1, \ldots, t_{i-2}) \otimes g_1(t_{i-1}) \right],$$
$$\left[N_1, \ldots, N_m \right] \left(I_m \otimes g_{i-1}(t_1, \ldots, t_{i-1}) \right) \Big], \quad i \geq 3.$$

Remark 1 We would like to mention that the $g_i(t_1, \ldots, t_i)$ are the Volterra kernels of the QB system (2) that map the input-to-output of the system. For more details on Volterra series expressions for QB systems, we refer to [13].

Furthermore, the connection between the above $\mathcal{H}_2$-norm definition for QB systems and the recently proposed algebraic Gramians for QB systems [12] has been studied in [13] as well. Therein, it is shown that the $\mathcal{H}_2$-norm can also be computed in terms of the Gramians as follows:

$$\|\Sigma\|_{\mathcal{H}_2} = \sqrt{\operatorname{tr}\left(C P C^T \right)} = \sqrt{\operatorname{tr}\left(B^T Q B \right)}, \tag{7}$$

where P and Q are the controllability and observability Gramians for QB systems which, respectively, satisfy the following quadratic Lyapunov equations:

$$AP + PA^T + H(P \otimes P)H^T + \sum_{k=1}^{m} N_k P N_k^T + BB^T = 0, \qquad (8)$$

$$A^T Q + QA + \mathscr{H}^{(2)}(P \otimes Q)\left(\mathscr{H}^{(2)}\right)^T + \sum_{k=1}^{m} N_k^T Q N_k + C^T C = 0. \qquad (9)$$

However, investigating an optimal model reduction problem for QB systems using the $\mathscr{H}_2$-norm is not a trivial task. Therefore, to ease the optimization problem, a concept of truncated $\mathscr{H}_2$-norms for QB systems has also been introduced in [13], which mainly relies on the first three terms of the corresponding Volterra series. Precisely, one possible truncated $\mathscr{H}_2$-norm $\|\Sigma\|_{\mathscr{H}_2^{\mathscr{T}}}$ is defined as follows:

$$\|\Sigma\|_{\mathscr{H}_2^{\mathscr{T}}} := = \sqrt{\mathrm{tr}\left(\sum_{i=1}^{3} \int_0^\infty \cdots \int_0^\infty \tilde{f}_i(t_1,\ldots,t_i)\tilde{f}_i(t_1,\ldots,t_i)^T dt_1 \cdots dt_i\right)}, \qquad (10)$$

in which

$$\tilde{f}_i(t_1,\ldots,t_i) = C\tilde{g}_i(t_1,\ldots,t_i),$$

where the $\tilde{g}_i(t_1,\ldots,t_i)$ satisfy

$$\tilde{g}_1(t_1) = e^{At_1} B,$$
$$\tilde{g}_2(t_1,t_2) = e^{At_2} \left[N_1, \ldots, N_m\right] (I_m \otimes \tilde{g}_1(t_1)),$$
$$\tilde{g}_3(t_1,t_2,t_3) = e^{At_3} H \left[\tilde{g}_1(t_1) \otimes \tilde{g}_1(t_2)\right].$$

Similar to the $\mathscr{H}_2$-norm expression, a connection between the truncated $\mathscr{H}_2$-norm (10) and the truncated Gramians for QB systems, introduced in [12], has been established in [13]. Thus, an alternative way to compute the truncated $\mathscr{H}_2$-norm is as follows:

$$\|\Sigma\|_{\mathscr{H}_2^{\mathscr{T}}} = \sqrt{\mathrm{tr}\left(C P_{\mathscr{T}} C^T\right)} = \sqrt{\mathrm{tr}\left(B^T Q_{\mathscr{T}} Q^T\right)},$$

where $P_{\mathscr{T}}$ are $Q_{\mathscr{T}}$ are, respectively, truncated versions of the controllability and observability Gramians, satisfying

$$A P_{\mathscr{T}} + P_{\mathscr{T}} A^T = -BB^T - H(P_1 \otimes P_1)H^T - \sum_{k=1}^{m} N_k P_1 N_k^T,$$

$$A^T Q_{\mathscr{T}} + Q_{\mathscr{T}} A = -C^T C - \mathscr{H}^{(2)}(P_1 \otimes Q_1)\left(\mathscr{H}^{(2)}\right)^T - \sum_{k=1}^{m} N_k^T Q_1 N_k,$$

in which P_1 and Q_1 are the unique solutions of the conventional Lyapunov equations:

$$AP_1 + P_1A^T = -BB^T, \qquad A^TQ_1 + Q_1A = -C^TC.$$

Using the truncated $\mathcal{H}_2$ measure, the aim is to construct a reduced-order system such that the measure of the error system is minimized. In other words, we need to determine a reduced-order system such that $\|\Sigma - \hat{\Sigma}\|_{\mathcal{H}_2^{\mathcal{T}}}$ is minimized. This problem has been studied in detail in [13], where, first, necessary conditions for optimality are derived. Based on these conditions, an iterative scheme is proposed, which upon convergence, yields a reduced-order system that *approximately* satisfies the derived optimality conditions. The iterative scheme is referred to as truncated quadratic-bilinear rational Krylov algorithm (TQB-IRKA). A brief sketch of TQB-IRKA is given in Algorithm 1 which considers a generalized nonsingular matrix E as well in the system (2).

Remark 2 To apply Algorithm 1, we need to compute Kronecker products such as $H(V_1 \otimes V_1)\tilde{H}^T$. In a large-scale setting, a direct computation of such Kronecker products is infeasible. As a remedy, in [9, 13], alternative methods are proposed to perform these computations efficiently by using some tools from tensor theory or by using the special structure of the Hessian H, arising from semi-discretization of PDEs.

Remark 3 In Algorithm 1, step 7, solving for V_1, is presented in a Sylvester equation form. However, using the vec $(\cdot)$ property (4), one can write this as a standard linear system, that is,

$$-(\Lambda \otimes E + I_r \otimes A)\,\mathrm{vec}\,(V_1) = \mathrm{vec}\left(B\tilde{B}^T\right), \qquad (12)$$

where I_r is the identity matrix of size $r \times r$. Similar expressions can also be derived for the matrices V_2, W_1, and W_2 in Algorithm 2. The above vec $(\cdot)$ form to compute the projection matrices is very useful for the latter part of the paper.

3 Transformation of a QBDAE into a QBODE and Model Reduction

Our next task is to investigate how TQB-IRKA (Algorithm 1) can be applied to QBDAEs, having the structure as in (1). For this, we first transform the system (1) into an equivalent ODE system by means of projections. Such a transformation is widely done in the literature for semi-discretized Navier–Stokes equations, see, e.g., [1, 21]. For completeness, we show the necessary steps that transform the system (1) into an equivalent ODE system. We begin by considering $B_2 = 0$ and the zero initial condition $v(0) = 0$ in (1), that is:

Algorithm 1: Truncated QB rational Krylov algorithm (TQB-IRKA) [13].

Input: The system matrices $E, A, H, N_1, \ldots, N_m, B, C$.
Output: Reduced matrices $\hat{E}, \hat{A}, \hat{H}, \hat{N}_1, \ldots, \hat{N}_m, \hat{B}, \hat{C}$.

1 Make an initial guess for the reduced matrices $\hat{E}, \hat{A}, \hat{H}, \hat{N}_1, \ldots, \hat{N}_m, \hat{B}, \hat{C}$.

2 **while** *not converged* **do**

3   Solve the generalized eigenvalue problem for the pencil $(\lambda \hat{E} - \hat{A})$, i.e., determine nonsingular matrices X and Y such that $X\hat{A}Y = \mathrm{diag}\,(\sigma_1, \ldots, \sigma_r) =: \Lambda$ and $X\hat{E}Y = I_r$ in which the σ_i's are the eigenvalues of the matrix pencil and I_r is the identity matrix of size $r \times r$.

4   Define $\tilde{H} = X\hat{H}(Y \otimes Y)$, $\tilde{N}_k = X\hat{N}_kY$, $\tilde{B} = X\hat{B}$ and $\tilde{C} = \hat{C}Y$.

5   Determine the mode-2 matricization of the tensor $\tilde{\mathscr{H}}$, denoted by $\tilde{\mathscr{H}}^{(2)}$ such that $\tilde{\mathscr{H}}^{(1)} = \tilde{H}$.

6   Solve for V_1 and V_2:

7       $-EV_1\Lambda - AV_1 = B\tilde{B}^T$,

8       $-EV_2\Lambda - AV_2 = H(V_1 \otimes V_1)\tilde{H}^T + \sum_{k=1}^m N_k V_1 \tilde{N}_k^T$.

9   Solve for W_1 and W_2:

10      $-E^T W_1\Lambda - A^T W_1 = C^T\tilde{C}$,

11      $-E^T W_2\Lambda - A^T W_2 = 2 \cdot \tilde{\mathscr{H}}^{(2)}(V_1 \otimes W_1)\left(\tilde{\mathscr{H}}^{(2)}\right)^T + \sum_{k=1}^m N_k^T W_1 \tilde{N}_k$.

12   Determine V and W:

13      $V := V_1 + V_2$ and $W := W_1 + W_2$.

14   Perform $V = \mathrm{orth}\,(V)$ and $W = \mathrm{orth}\,(W)$.

15   Compute reduced-order matrices:

16      $\hat{E} = W^T EV$, $\hat{A} = W^T AV$, $\hat{N}_k = W^T N_k V$, $k \in \{1, \ldots, m\}$,

17      $\hat{H} = W^T H(V \otimes V)$, $\hat{B} = W^T B$, $\hat{C} = CV$.

18 **end**

$$E_{11}\dot{v}(t) = A_{11}v(t) + A_{12}p(t) + Hv(t) \otimes v(t) + \sum_{k=1}^m N_k v(t)u_k(t) + B_1 u(t), \quad (13a)$$

$$0 = A_{21}v(t), \quad v(0) = 0, \tag{13b}$$

$$y(t) = C_1 v(t) + C_2 p(t). \tag{13c}$$

From (13b), we get $A_{21}\frac{d}{dt}v(t) = 0$, leading to

$$0 = A_{21}\frac{d}{dt}v(t) = A_{21}E_{11}^{-1}\left(E_{11}\frac{d}{dt}v(t)\right)$$

$$= A_{21}E_{11}^{-11}\left(A_{11}v(t) + A_{12}p(t) + Hv(t) \otimes v(t) + \sum_{k=1}^m N_k v(t)u_k(t) + B_1 u(t)\right)$$

(using (13a)).

As a result, we obtain an explicit expression for the pressure $p(t)$ as

$$p(t) = -S^{-1}A_{21}E_{11}^{-1}\left(A_{11}v(t) + Hv(t) \otimes v(t) + \sum_{k=1}^{m}N_k v(t)u_k(t) + B_1 u(t)\right), \quad (14)$$

where $S = A_{21}E_{11}^{-1}A_{12}$. Substituting for $p(t)$ using (14) in (13a) and (13c) yields

$$E_{11}\dot{v}(t) = \Pi_l A_{11}v(t) + \Pi_l Hv(t) \otimes v(t) + \sum_{k=1}^{m}\Pi_l N_k v(t)u_k(t) + \Pi_l B_1 u(t), \quad (15a)$$

$$y(t) = \mathscr{C}v(t) + \mathscr{C}_H v(t) \otimes v(t) + \sum_{k=1}^{m}\mathscr{C}_{N_k}v(t)u_k(t) + \mathscr{D}u(t), \quad v(0) = 0. \quad (15b)$$

where

$$\mathscr{C} = C_1 - C_2 S^{-1}A_{21}E_{11}^{-1}A_{11}, \qquad \mathscr{C}_H = -C_2 S^{-1}A_{21}E_{11}^{-1}H,$$
$$\mathscr{C}_{N_k} = -C_2 S^{-1}A_{21}E_{11}^{-1}N_k, \qquad \mathscr{D} = -C_2 S^{-1}A_{21}E_{11}^{-1}B_1, \quad \text{and}$$
$$\Pi_l = I_{n_v} - A_{12}S^{-1}A_{21}E_{11}^{-1}.$$

Recall the properties of Π_l from [21], which are as follows:

$$\Pi_l^2 = \Pi_l, \qquad \text{range}(\Pi_l) = \ker\left(A_{21}E_{11}^{-1}\right), \qquad \ker(\Pi_l) = \text{range}(A_{12}).$$

This implies that Π_l is an *oblique* projector. Moreover, for later purpose, we also define another oblique projector

$$\Pi_r = I_{n_v} - E_{11}^{-1}A_{12}S^{-1}A_{21}.$$

First, we note a relation between Π_l and Π_r, that is, $\Pi_l E_{11} = E_{11}\Pi_r$. Moreover, it can be verified that

$$A_{21}z = 0 \quad \text{if and only if} \quad \Pi_r z = z.$$

Using the properties of the projectors Π_l and Π_r, we obtain

$$\Pi_l E_{11}\Pi_r \dot{v}(t) = \Pi_l A_{11}\Pi_r v(t) + \Pi_l H\left(\Pi_r v(t) \otimes \Pi_r v(t)\right) + \sum_{k=1}^{m}\Pi_l N_k \Pi_r v(t)u_k(t)$$
$$+ \Pi_l B_1 u(t), \quad (16a)$$

$$y(t) = \mathscr{C}\Pi_r v(t) + \mathscr{C}_H \Pi_r v(t) \otimes \Pi_r v(t) + \sum_{k=1}^{m}\mathscr{C}_{N_k}\Pi_r v(t)u_k(t) + \mathscr{D}u(t), \quad (16b)$$

with zero initial condition, i.e., $v(0) = 0$. Moreover, the dynamics of the system (16) lies in an $n_v - n_p$ dimensional subspace, that is nothing but the null space of Π_l. Next, we decompose the projectors Π_l and Π_r as follows:

$$\Pi_l = \theta_l \phi_l^T, \qquad \Pi_r = \theta_r \phi_r^T, \tag{17}$$

in which $\theta_i, \phi_i \in \mathbb{R}^{n_v \times n_v - n_p}$ for $i \in \{l, r\}$ satisfy

$$\theta_l^T \phi_l = I_{n_v - n_p}, \qquad \theta_r^T \phi_r = I_{n_v - n_p}.$$

Thus, if one defines $\tilde{v}(t) := \phi_r^T v(t)$, we consequently obtain an equivalent ODE system of the system (16) as follows:

$$\phi_l^T E_{11} \theta_r \dot{\tilde{v}}(t) = \phi_l^T A_{11} \theta_r \tilde{v}(t) + \phi_l^T H \theta_r \tilde{v}(t) \otimes \theta_r \tilde{v}(t) + \sum_{k=1}^{m} \phi_l^T N_k \theta_r \tilde{v}(t) u_k(t)$$
$$+ \phi_l^T B_1 u(t), \tag{18a}$$

$$y(t) = \mathscr{C} \theta_r \tilde{v}(t) + \mathscr{C}_H \theta_r \tilde{v}(t) \otimes \theta_r \tilde{v}(t) + \sum_{k=1}^{m} \mathscr{C}_{N_k} \theta_r \tilde{v}(t) u_k(t) + \mathscr{D} u(t), \tag{18b}$$

with $\tilde{v}(0) = 0$. Thus, one can apply Algorithm 1 to obtain projection matrices which give a *near*-optimal reduced system, having neglected nonlinear terms in the output Eq. (18b). However, there are two major issues: this requires to compute ϕ_l, θ_r, or Π_l, Π_r which are expensive to compute, and in case we are able to determine these projectors and their decompositions efficiently, the matrix coefficients of the system (18), e.g., $\phi_l^T E_{11} \theta_r$, $\phi_l^T A_{11} \theta_r$, might be dense, thus making MOR techniques numerically inefficient and expensive. Therefore, in the following, we discuss how to employ Algorithm 1 to the system (18) without requiring explicit computation of the projection matrices and their decompositions.

Computational Issues

In order to determine the projection matrices to compute a reduced system, we consider the following associated QB system, which is the system (18), having neglected the nonlinear terms in the output equation:

$$\bar{E} \dot{\tilde{v}}(t) = \bar{A} \tilde{v}(t) + \bar{H} \tilde{v}(t) \otimes \tilde{v}(t) + \sum_{k=1}^{m} \bar{N}_k \tilde{v}(t) u_k(t) + \bar{B}_1 u(t), \tag{19a}$$

$$\tilde{y}(t) = \bar{C} \theta_r \tilde{v}(t) + \mathscr{D} u(t), \quad \tilde{v}(0) = 0, \tag{19b}$$

where

$$\bar{E} := \phi_l^T E_{11} \theta_r, \qquad \bar{A} := \phi_l^T A_{11} \theta_r, \quad \bar{N}_k := \phi_l^T N_k \theta_r, \quad k \in \{1, \ldots, m\},$$
$$\bar{H} := \phi_l^T H \theta_r \otimes \theta_r, \qquad \bar{B}_1 := \phi_l^T B_1, \qquad \bar{C} := \mathscr{C} \theta_r. \tag{20}$$

Here, we need to neglect the nonlinear terms appearing in the output equation due to the transformation to an ODE system as Algorithm 1 is developed so far only for linear output equations. Thus, the full incorporation of the nonlinearities in the output equation requires further work. However, note that the nonlinear terms are again included in the reduced-order system by projecting the nonlinear terms in the output equation of (18) using the projection matrices obtained by TQB-IRKA applied to the system (19). This will be discussed later in Remark 4 in detail.

As a first step, we aim at determining the projection matrices V and W such that the original matrices like E_{11}, A_{11} can be used to compute the reduced matrices instead of using, e.g., $\bar{E}$, $\bar{A}$, i.e., a reduced matrix $\hat{E}$ can be computed as $W^T E_{11} V$, and so on.

For this purpose, let $\bar{V}$ and $\bar{W}$ be the solutions of the Sylvester equations in Algorithm 1 (steps 6–11) using the matrices $\bar{E}$, $\bar{A}$, etc. Furthermore, we define the matrices V and W, satisfying:

$$V = \theta_r \bar{V}, \qquad \text{and} \qquad W = \phi_l \bar{W}. \tag{21}$$

Then, it can easily be verified that the reduced matrices computed using the quantities marked with *bar*, e.g., $\bar{V}$, $\bar{W}$, $\bar{E}$, are the same as the reduced matrices computed using V and W and the original matrices. In other words, $\bar{W}^T \bar{E} \bar{V} = W^T E_{11} V$ and so forth. Using this formulation, we sketch an algorithm for model reduction of (19) which gives near-optimal reduced systems on convergence by projecting the original system matrices, see Algorithm 2.

However, to compute the projection matrices at each iteration in Algorithm 2 (steps 7–12), we still require the basis matrices ϕ_l and θ_r. Therefore, our next goal is to determine these projection matrices, without involving ϕ_l and θ_r, or more precisely, we aim at computing the matrices V and W using only the original matrices such as E_{11}, A_{11}, A_{12}.

Luckily, a similar problem has been studied in [17], where for the symmetric case, i.e., $\Pi_l = \Pi_r^T$, it is shown how to compute $(I_r \otimes \theta_r)L(I_r \otimes \phi_l^T)f$ for a given arbitrary vector f and the matrix L at step 6 in Algorithm 2, without explicitly forming θ_r and ϕ_l. For the unsymmetric case, i.e., $\Pi_l \neq \Pi_r^T$, the following result can be developed in a similar fashion as in [17], and, therefore, a detailed proof is omitted.

Lemma 1 *Consider ϕ_l and θ_r as in (17), and assume $\mathscr{X} = (I_r \otimes \phi_l^T)\mathscr{T}(I_r \otimes \theta_r)$ is invertible for a given $\mathscr{T}$. Furthermore, let $\mathscr{G}$ and $\mathscr{G}_T$ be $(I_r \otimes \theta_r)\mathscr{X}^{-1}(I_r \otimes \phi_l^T)$ and $(I_r \otimes \theta_l)\mathscr{X}^{-T}(I_r \otimes \phi_r^T)$, respectively. Then, the vector*

$$\bar{v} = \mathscr{G} f$$

Algorithm 2: Truncated QB rational Krylov algorithm for the system (19), involving projectors.

Input: The system matrices $E_{11}, A_{11}, H, N_1, \ldots, N_m, B_1, \mathscr{C}$.
Output: Redcued matrices $\hat{E}, \hat{A}, \hat{H}, \hat{N}_1, \ldots, \hat{N}_m, \hat{B}, \hat{C}$.

1 Make an initial guess of reduced matrices $\hat{E}, \hat{A}, \hat{H}, \hat{N}_1, \ldots, \hat{N}_m, \hat{B}, \hat{C}$.
2 **while** *not converged* **do**
3   Solve the generalized eigenvalue problem for the pencil $(\lambda \hat{E} - \hat{A})$, i.e., determine nonsingular matrices X and Y such that $X \hat{A} Y = \text{diag}(\sigma_1, \ldots, \sigma_r) =: \Lambda$ and $X \hat{E} Y = I_r$ in which the σ_i's are the eigenvalues of the matrix pencil.
4   Define $\tilde{H} = X \hat{H}(Y \otimes Y)$, $\tilde{N}_k = X \hat{N}_k Y$, $\tilde{B} = X \hat{B}$ and $\tilde{C} = \hat{C} Y$.
5   Determine the mode-2 matricization of the tensor $\tilde{\mathscr{H}}$, denoted by $\tilde{\mathscr{H}}^{(2)}$ such that $\tilde{\mathscr{H}}^{(1)} = \tilde{H}$.
6   Define $L := \left(-\Lambda \otimes \bar{E} - I_r \otimes \bar{A} \right)^{-1}$, where $\bar{E}$ and $\bar{A}$ are as in (20).
7   Solve for V_1 and V_2:
8         $\text{vec}(V_1) = (I_r \otimes \theta_r) L (I_r \otimes \phi_l^T) \text{vec} \left(B \tilde{B}^T \right)$,
9         $\text{vec}(V_2) = (I_r \otimes \theta_r) L (I_r \otimes \phi_l^T) \text{vec} \left(H(V_1 \otimes V_1) \tilde{H}^T + \sum_{k=1}^m N_k V_1 \tilde{N}_k^T \right)$.
10   Solve for W_1 and W_2:
11         $\text{vec}(W_1) = (I_r \otimes \phi_l) L^T (I_r \otimes \theta_r^T) \text{vec} \left(\mathscr{C}^T \tilde{C} \right)$,
12         $\text{vec}(W_2) =$
        $(I_r \otimes \phi_l) L^T (I_r \otimes \theta_r^T) \text{vec} \left(2 \cdot \mathscr{H}^{(2)} (V_1 \otimes W_1) \left(\tilde{\mathscr{H}}^{(2)} \right)^T + \sum_{k=1}^m N_k^T W_1 \tilde{N}_k \right)$.
13   Determine V and W:
14     $V := V_1 + V_2$ and $W := W_1 + W_2$.
15   Perform $V = \text{orth}(V)$ and $W = \text{orth}(W)$.
16   Compute reduced-order matrices:
17     $\hat{E} = W^T E_{11} V$, $\hat{A} = W^T A_{11} V$, $\hat{N}_k = W^T N_k V$, $k \in \{1, \ldots, m\}$,
18     $\hat{H} = W^T H(V \otimes V)$, $\hat{B} = W^T B_1$, $\hat{C} = \mathscr{C} V$.
19 **end**

---

solves

$$\begin{bmatrix} \mathscr{T} & I_r \otimes A_{12} \\ I_r \otimes A_{21} & 0 \end{bmatrix} \begin{bmatrix} \bar{v} \\ \xi_v \end{bmatrix} = \begin{bmatrix} f \\ 0 \end{bmatrix}.$$

Similarly, the vector

$$\bar{w} = \mathscr{G}_T g$$

solves

$$\begin{bmatrix} \mathscr{T}^T & I_r \otimes A_{21}^T \\ I_r \otimes A_{12}^T & 0 \end{bmatrix} \begin{bmatrix} \bar{w} \\ \xi_w \end{bmatrix} = \begin{bmatrix} g \\ 0 \end{bmatrix}.$$

By making use of the result of Lemma 1, it can be shown that the steps 7–12 in Algorithm 2 can be performed without an explicit requirement of the basis matrices ϕ_l and θ_r. We rather need to solve appropriate saddle point problems to compute V_i

---

Algorithm 3: Truncated QB rational Krylov algorithm for the system (19).

Input: The system matrices $E_{11}, A_{11}, H, N_1, \ldots, N_m, B_1, \mathscr{C}$.

Output: Redcued matrices $\hat{E}, \hat{A}, \hat{H}, \hat{N}_1, \ldots, \hat{N}_m, \hat{B}, \hat{C}$.

1 Make an initial guess of reduced matrices $\hat{E}, \hat{A}, \hat{H}, \hat{N}_1, \ldots, \hat{N}_m, \hat{B}, \hat{C}$.

2 **while** *not converged* **do**

3 $\quad$ Solve the generalized eigenvalue problem for the pencil $(\lambda \hat{E} - \hat{A})$, i.e., determine nonsingular matrices X and Y such that $X \hat{A} Y = \text{diag}(\sigma_1, \ldots, \sigma_r) =: \Lambda$ and $X \hat{E} Y = I_r$ in which the σ_i's are the eigenvalues of the matrix pencil.

4 $\quad$ Define $\tilde{H} = X \hat{H}(Y \otimes Y)$, $\tilde{N}_k = X \hat{N}_k Y$, $\tilde{B} = X \hat{B}$ and $\tilde{C} = \hat{C} Y$.

5 $\quad$ Determine the mode-2 matricization of the tensor $\tilde{\mathscr{H}}$, denoted by $\tilde{\mathscr{H}}^{(2)}$, such that $\tilde{\mathscr{H}}^{(1)} = \tilde{H}$.

6 $\quad$ Define $L := \begin{bmatrix} (-\Lambda \otimes E_{11} - I_r \otimes A_{11})^{-1} & I_r \otimes A_{12} \\ I_r \otimes A_{21} & 0 \end{bmatrix}$, and $\mathscr{I} = \begin{bmatrix} I_{n_v} & 0 \end{bmatrix}$.

7 $\quad$ Solve for V_1 and V_2:

8 $\quad\quad \text{vec}(V_1) = \mathscr{I} L \begin{bmatrix} \text{vec}\left(B \tilde{B}^T \right) \\ 0 \end{bmatrix}$,

9 $\quad\quad \text{vec}(V_2) = \mathscr{I} L \begin{bmatrix} \text{vec}\left(H(V_1 \otimes V_1) \tilde{H}^T + \sum_{k=1}^{m} N_k V_1 \tilde{N}_k^T \right) \\ 0 \end{bmatrix}$.

10 $\quad$ Solve for W_1 and W_2:

11 $\quad\quad \text{vec}(W_1) = \mathscr{I} L^T \begin{bmatrix} \text{vec}\left(\mathscr{C}^T \tilde{C} \right) \\ 0 \end{bmatrix}$,

12 $\quad\quad \text{vec}(W_2) = \mathscr{I} L^T \begin{bmatrix} \text{vec}\left(2 \cdot \mathscr{H}^{(2)}(V_1 \otimes W_1) \left(\tilde{\mathscr{H}}^{(2)} \right)^T + \sum_{k=1}^{m} N_k^T W_1 \tilde{N}_k \right) \\ 0 \end{bmatrix}$.

13 $\quad$ Determine V and W:

14 $\quad\quad V := V_1 + V_2$ and $W := W_1 + W_2$.

15 $\quad$ Perform $V = \text{orth}(V)$ and $W = \text{orth}(W)$.

16 $\quad$ Compute reduced-order matrices:

17 $\quad\quad \hat{E} = W^T E_{11} V, \quad\quad \hat{A} = W^T A_{11} V, \quad\quad \hat{N}_k = W^T N_k V, \quad k \in \{1, \ldots, m\},$

18 $\quad\quad \hat{H} = W^T H(V \otimes V), \quad \hat{B} = W^T B_1, \quad\quad \hat{C} = \mathscr{C} V.$

19 **end**

---

and W_i, $i \in \{1, 2\}$. All these analyses lead to Algorithm 3 that does not require any explicit computation of the basis matrices ϕ_l and θ_r.

Remark 4 Recall that Algorithm 3 upon convergence gives a *near*-optimal reduced system for the system (18), having neglected the nonlinear terms in the output equation in the model reduction process. Nonetheless, we reduce these nonlinear terms in the output equation using the projection matrix V, obtained upon convergence, i.e., $\hat{\mathscr{C}}_H = \mathscr{C}_H(V \otimes V)$ and $\hat{\mathscr{C}}_{N_k} = \mathscr{C}_{N_k} V$. Furthermore, if the output of the system (13) is given only by linear combinations of the velocity $v(t)$, then all nonlinear terms in the output equation of the system (18) are zero.

Remark 5 Throughout the above discussion, we have assumed that $B_2 = 0$ in (1). However, as discussed, e.g., in [21], the general case $B_2 \neq 0$ can be converted into the case $B_2 = 0$ by an appropriate change of variables for $v(t)$. For this, one needs to decompose $v(t)$ as follows:

$$v(t) = v_m(t) + v_u(t), \tag{22}$$

where $v_u(t) = \Omega u(t)$ in which $\Omega := -E_{11}^{-1}A_{12}(A_{21}E_{11}A_{12})B_2$. If one substitutes for $v(t)$ from (22) into (1), then it can be easily seen that $A_{21}v_m(t) = 0$. Furthermore, we assume the initial condition to be consistent. Performing similar algebraic calculations as done for the problem $B_2 = 0$ leads to the following system:

$$\Pi_l E_{11} \Pi_r \dot{v}(t) = \Pi_l A_{11} \Pi_r v(t) + \Pi_l H (\Pi_r v(t) \otimes \Pi_r v(t)) + \sum_{k=1}^{m} \Pi_l \mathcal{N}_k \Pi_r v(t) u_k(t)$$
$$+ \Pi_l \mathcal{B}_1 u(t) + \Pi_l \mathcal{B}_u (u(t) \otimes u(t)), \tag{23a}$$

$$y(t) = \mathcal{C} v(t) + \mathcal{C}_H \Pi_r v(t) \otimes \Pi_r v(t) + \sum_{k=1}^{m} \mathcal{C}_{N_k} \Pi_r v(t) u_k(t) + \mathcal{D} u(t)$$
$$- C_2 (A_{21} E_{11}^{-1} B_2) \dot{u}(t), \tag{23b}$$

where

$$\mathcal{N}_k = N_k - H(I \otimes \Omega_k + \Omega_k \otimes I), \qquad \mathcal{B}_1 = B_1 + A_{11}\Omega,$$
$$\mathcal{B}_u = H(\Omega \otimes \Omega) + [N_1\Omega, \ldots, N_m\Omega], \qquad \mathcal{C} = C_1 - C_2 S^{-1} A_{21} E_{11}^{-1} A_{11},$$
$$\mathcal{C}_H = -C_2 S^{-1} A_{21} E_{11}^{-1} H, \qquad \mathcal{C}_{N_k} = -C_2 S^{-1} A_{21} E_{11}^{-1} \mathcal{N}_k,$$

in which $S := A_{21} E_{11}^{-1} A_{12}$. Thus, we obtain a system equivalent to (1) which has a similar form as in (16). However, the system (23) contains some extra terms which are functions of the input $u(t)$, e.g., $u_k \cdot u_q$, $\{k, q\} \in \{1, \ldots, m\}$ and the derivative of the input $u(t)$. Although they are functions of the input in a forward simulation, we consider them formally as different inputs as far as the model reduction problem is concerned. Hence, Algorithm 3 can readily be applied to the system (23) to determine reduced-order systems.

4 Numerical Experiment

In this section, we test the efficiency of Algorithm 3 using as numerical example the lid-driven cavity, obtained using semi-discretization of the Navier–Stokes equations. We initialize Algorithm 3 randomly and iterate it until the relative change in the eigenvalues of the reduced matrix $\hat{A}$ is less than 10^{-4}. All the simulations were done on a board with 4 Intel® Xeon® E7-8837 CPUs with a 2.67-GHz clock speed using MATLAB® 8.0.0.783 (R2012b).

Lid-Driven Cavity

Here, we consider a lid-driven cavity as shown in Fig. 1, which is modeled using the incompressible Navier–Stokes equations in the velocity $\breve{v}$ and the pressure $\breve{p}$ on the unit square $\Omega = (0, 1)^2$ and boundary Γ. The governing equations are:

$$\dot{\breve{v}} + (\breve{v} \cdot \nabla) \, \breve{v} - \frac{1}{\mathrm{Re}} \Delta \breve{v} + \nabla \breve{p} = 0, \tag{24a}$$

$$\nabla \cdot \breve{v} = 0, \tag{24b}$$

and boundary and initial conditions:

$$\breve{v}|_\Gamma = g, \qquad \breve{v}|_{t=0} = \breve{v}_0,$$

where Re is the Reynolds number, g represents the Dirichlet boundary conditions, we set $\breve{v} = \begin{bmatrix} 1 \\ 0 \end{bmatrix}$ for the upper boundary and zero at the other boundaries, and $\breve{v}_0$ is an initial condition. Furthermore, the system is subject to a control input $u(t)$ in the domain Ω_c. As a result, we ensure that the x and y components of the velocity in the domain Ω_c are $u(t)$.

Having applied a finite element scheme using the Taylor–Hood scheme on a uniform mesh, a discretized system in the velocity v and pressure p is obtained as follows:

$$E_{11}\dot{v} = A_{11}v(t) + H(v \otimes v) + A_{12}v(t) + f + B_1 u(t), \tag{25a}$$

$$0 = A_{12}^T v(t), \qquad v(0) = v_s, \tag{25b}$$

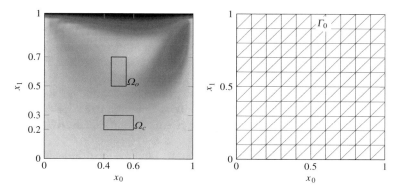

Fig. 1 Illustration of the velocity magnitude of the lid-driven cavity problem and the domain of control and observation $\Omega_c = [0.4, 0.6] \times [0.2, 0.3]$ and $\Omega_o = [0.45, 0.55] \times [0.5, 0.7]$. Cf. [5]

where $E_{11}, A_{11} \in \mathbb{R}^{n_v \times n_v}, A_{12} \in \mathbb{R}^{n_v \times n_p}, H \in \mathbb{R}^{n_v \times n_v^2}, f \in \mathbb{R}^{n_v}$, and v_s is the steady-state solution of the system when the control input u is set to zero. Furthermore, we observe the average velocity in x and y directions in the domain Ω_o, leading to an output equation as follows:

$$y(t) = Cv(t). \tag{26}$$

For a detailed description of the weak formulation of the Navier–Stokes equations with a distributed input, leading to the state-space representation (25) from (24) for the lid-driven cavity example, we refer to [5]. Therein, a detailed discussion of the observation domain, more preciously of the output Eq. (26) can also be found.

Furthermore, we assume an initial condition to be the steady-state solution of the system; thus, we perform a change of variables as $v_\delta = v + v_s$ and $p_\delta = p + p_s$ to ensure the zero initial condition of the transformed system. This results in the following system:

$$E_{11}\dot{v}_\delta = (A_{11} + X)\, v_\delta(t) + H(v_\delta \otimes v_\delta) + A_{12}p_\delta(t) + B_1 u(t),$$
$$0 = A_{12}^T v_\delta(t), \qquad v_\delta(0) = 0,$$
$$y(t) = Cv_\delta + D,$$

where $X := H(v_s \otimes I + I \otimes v_s)$ and $D := Cv_s$. We obtain the degrees of freedom for the velocity and the pressure to be 3042 and 440, respectively, i.e., $n_v = 3042$ and $n_p = 440$. We set the Reynolds number Re $= 100$. Furthermore, we select four cells in the domain Ω_o on which we measure the average x and y components of the velocity, thus have eight outputs. Next, we employ Algorithm 3 to obtain a reduced-order system of order $r = 140$. We would like to mention that the order of the reduced system is chosen by gradually increasing it until a satisfactory reduced system is obtained, and therefore, it is a potentially future research topic to derive a priori error estimation, allowing us to choose a priori an appropriate order of the reduced system.

To check the accuracy of the reduced system, we perform time-domain simulations for the original and reduced systems for an arbitrary control input $u(t) = 2t^2 \exp(-t/2) \sin(2\pi t/5)$ and observe the outputs which are plotted in Fig. 2. This shows that the reduced system captures the dynamics of the system faithfully. Furthermore, we also plot the velocity at the full grid in Fig. 3, although the considered model reduction aims at capturing only the input-to-output behavior of the system. The figure indicates that the reduced-order system can replicate the dynamics of the original system at the full grid as well.

Lastly, we compare the CPU time required to simulate the original and reduced-order systems. Since the original system is of descriptor nature, we simulate it with an efficiently implemented implicit Euler method with a uniform time step of 0.05 s. On the other side, one of the main advantages of the reduced-order system being an ODE system is that it can be easily simulated by using an ODE solver in MATLAB®, and thus we choose ode15s to simulate the reduced system in the time interval $(0, 10]$. We observe that the original system requires 303.38 s, whereas the reduced-order

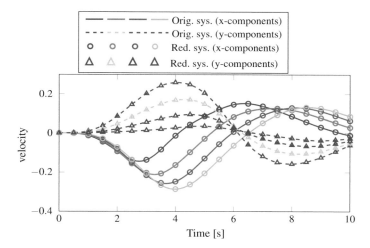

Fig. 2 For a random control $u(t) = 2t^2 \exp(-t/2) \sin(2\pi t/5)$, we compare the average x and y components of velocities at four cells in the domain Ω_o obtained via the original system ($n = 3482$) and reduced system ($r = 140$) in the figure

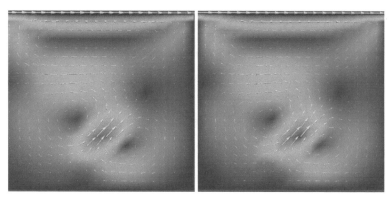

Fig. 3 Comparison of $|v|$ obtained from the original (left) and reduced (right) systems at time $t = 3.75\,\text{s}$ for an input $u(t) = 2t^2 \exp(-t/2) \sin(2\pi t/5)$

system needs 7.23 s which is nearly 40 times faster in our preliminary implementation. However, we point out that there is a huge potential to speed up the simulations of the original and reduced systems, e.g., several softwares and packages, specialized for flow problems, can be used for the original systems, and a sophisticated tensor technique can also be utilized to speed up simulations of the reduced-order systems.

5   Conclusions

In this work, we have studied the model reduction problem for a particular class of quadratic-bilinear descriptor systems, especially arising from the semi-discretization of the Navier–Stokes equations. In particular, we have investigated how one can employ the iterative model reduction scheme proposed in [13] for quadratic-bilinear ODEs to the latter class of descriptor systems. To that end, we have first transformed the quadratic-bilinear descriptor system into an equivalent ODE system by means of projectors. Furthermore, for practical computations, we have proposed an efficient iterative scheme for the considered quadratic-bilinear descriptor systems to avoid the explicit computation of the projectors that are used to transform the system into an ODE system. The advantage of this scheme over common model reduction techniques for flow equations such as POD is that the structure of the full-order system is preserved, and the reduced-order model is therefore independent of the input signal which can thus be varied in the reduced model without loss of accuracy. This important feature for applications in flow control will be studied further in the future. Using as numerical example a semi-discretized Navier–Stokes equation, we have shown the efficiency of the proposed method.

As a further research topic, an extension of the balanced truncation method for quadratic-bilinear ODE systems [12] to descriptor systems will be very useful. Furthermore, it will be a significant contribution to extend these methods to other classes of descriptor systems, particularly, appearing in mechanical systems.

Acknowledgements The authors would like to thank Dr. Jan Heiland for providing the lid-driven cavity model, used in the numerical experiment.

References

1. Ahmad, M.I., Benner, P., Goyal, P., Heiland, J.: Moment-matching based model reduction for Navier-Stokes type quadratic-bilinear descriptor systems. Z. Angew. Math. Mech (2017)
2. Antoulas, A.C.: Approximation of Large-Scale Dynamical Systems. SIAM Publications, Philadelphia, PA (2005)
3. Astrid, P., Weiland, S., Willcox, K., Backx, T.: Missing point estimation in models described by proper orthogonal decomposition. IEEE Trans. Autom. Control $53(10)$, 2237–2251 (2008)
4. Baur, U., Benner, P., Feng, L.: Model order reduction for linear and nonlinear systems: a system-theoretic perspective. Arch. Comput. Methods Eng. $21(4)$, 331–358 (2014)
5. Behr, M., Benner, P., Heiland, J.: Example setups of Navier-Stokes equations with control and observation: Spatial discretization and representation via linear-quadratic matrix coefficients (2017). arXiv:1707.08711
6. Benner, P., Breiten, T.: Interpolation-based $\mathcal{H}_2$-model reduction of bilinear control systems. SIAM J. Matrix Anal. Appl. $33(3)$, 859–885 (2012)
7. Benner, P., Breiten, T.: Krylov-subspace based model reduction of nonlinear circuit models using bilinear and quadratic-linear approximations. In: Günther, M., Bartel, A., Brunk, M., Schöps, S., Striebel, M. (eds.) Progress in Industrial Mathematics at ECMI 2010. Mathematics in Industry, vol. 17, pp. 153–159. Springer-Verlag, Berlin (2012)

8. Benner, P., Breiten, T.: Two-sided moment matching methods for nonlinear model reduction. Preprint MPIMD/12-12, MPI Magdeburg (2012). http://www.mpi-magdeburg.mpg.de/preprints/
9. Benner, P., Breiten, T.: Two-sided projection methods for nonlinear model reduction. SIAM J. Sci. Comput. **37**(2), B239–B260 (2015)
10. Benner, P., Damm, T.: Lyapunov equations, energy functionals, and model order reduction of bilinear and stochastic systems. SIAM J. Control Optim. **49**(2), 686–711 (2011)
11. Benner, P., Goyal, P.: Multipoint interpolation of Volterra series and $\mathcal{H}_2$-model reduction for a family of bilinear descriptor systems. Syst. Control Lett. **97**, 1–11 (2016)
12. Benner, P., Goyal, P.: Balanced truncation model order reduction for quadratic-bilinear control systems (2017). arXiv:1705.00160
13. Benner, P., Goyal, P., Gugercin, S.: $\mathcal{H}_2$-quasi-optimal model order reduction for quadratic-bilinear control systems (2016). arXiv:1610.03279
14. Benner, P., Mehrmann, V., Sorensen, D.C.: Dimension Reduction of Large-Scale Systems, vol. 45. Lecture Notes in Computational Science and Engineering. Springer, Heidelberg (2005)
15. Chaturantabut, S., Sorensen, D.C.: Nonlinear model reduction via discrete empirical interpolation. SIAM J. Sci. Comput. **32**(5), 2737–2764 (2010)
16. Flagg, G., Gugercin, S.: Multipoint Volterra series interpolation and $\mathcal{H}_2$ optimal model reduction of bilinear systems. SIAM J. Matrix Anal. Appl. **36**(2), 549–579 (2015)
17. Goyal, P., Benner, P.: An iterative model order reduction scheme for a special class of bilinear descriptor systems appearing in constraint circuit simulation. In: ECCOMAS Congress 2016, VII European Congress on Computational Methods in Applied Sciences and Engineering, vol. 2, pp. 4196–4212 (2016)
18. Gu, C.: QLMOR: a projection-based nonlinear model order reduction approach using quadratic-linear representation of nonlinear systems. IEEE Trans. Comput. Aided Des. Integr. Circuits Syst. **30**(9), 1307–1320 (2011)
19. Gugercin, S., Stykel, T., Wyatt, S.: Model reduction of descriptor systems by interpolatory projection methods. SIAM J. Sci. Comput. **35**(5), B1010–B1033 (2013)
20. Hairer, E., Wanner, G.: Solving Ordinary Differential Equations II-Stiff and Differential-Algebraic Problems, 2nd edn. Springer Series in Computational Mathematics. Springer (2002)
21. Heinkenschloss, M., Sorensen, D.C., Sun, K.: Balanced truncation model reduction for a class of descriptor systems with applications to the Oseen equations. SIAM J. Sci. Comput. **30**(2), 1038–1063 (2008)
22. Hinze, M., Volkwein, S.: Proper orthogonal decomposition surrogate models for nonlinear dynamical systems: error estimates and suboptimal control. In: [14], pp. 261–306
23. Kolda, T.G., Bader, B.W.: Tensor decompositions and applications. SIAM Rev. **51**(3), 455–500 (2009)
24. Kunisch, K., Volkwein, S.: Galerkin proper orthogonal decomposition methods for a general equation in fluid dynamics. SIAM J. Numer. Anal. **40**(2), 492–515 (2002)
25. Kunisch, K., Volkwein, S.: Proper orthogonal decomposition for optimality systems. ESAIM Math. Model. Numer. Anal. **42**(1), 1–23 (2008)
26. Schilders, W.H.A., van der Vorst, H.A., Rommes, J.: Model Order Reduction: Theory, Research Aspects and Applications. Springer, Heidelberg (2008)
27. Zhang, L., Lam, J.: On H_2 model reduction of bilinear systems. Automatica **38**(2), 205–216 (2002)

Greedy Kernel Approximation for Sparse Surrogate Modeling

Bernard Haasdonk and Gabriele Santin

Abstract Modern simulation scenarios frequently require multi-query or real-time responses of simulation models for statistical analysis, optimization, or process control. However, the underlying simulation models may be very time-consuming rendering the simulation task difficult or infeasible. This motivates the need for rapidly computable surrogate models. We address the case of surrogate modeling of functions from vectorial input to vectorial output spaces. These appear, for instance, in simulation of coupled models or in the case of approximating general input–output maps. We review some recent methods and theoretical results in the field of greedy kernel approximation schemes. In particular, we recall the vectorial kernel orthogonal greedy algorithm (VKOGA) for approximating vector-valued functions. We collect some recent convergence statements that provide sound foundation for these algorithms, in particular quasi-optimal convergence rates in case of kernels inducing Sobolev spaces. We provide some initial experiments that can be obtained with non-symmetric greedy kernel approximation schemes. The results indicate better stability and overall more accurate models in situations where the input data locations are not equally distributed.

1 Introduction

During the last decades, scientific computing has obtained large success in accurate simulation of complex natural processes, for example, from physics, chemistry, biology, or finance. Those models frequently depend on partial differential equations, which are discretized and then result in very large dimensional models. In other cases, a model consists of several submodels, which are coupled. Modern numerical techniques such as finite elements, adaptivity in space and time, and computer hardware

B. Haasdonk · G. Santin (✉)
Institute of Applied Analysis and Numerical Simulation, University of Stuttgart,
Pfaffenwaldring 57, 70569 Stuttgart, Germany
e-mail: santinge@mathematik.uni-stuttgart.de

B. Haasdonk
e-mail: haasdonk@mathematik.uni-stuttgart.de

© Springer International Publishing AG, part of Springer Nature 2018   21
W. Keiper et al. (eds.), *Reduced-Order Modeling (ROM) for Simulation and Optimization*,
https://doi.org/10.1007/978-3-319-75319-5_2

and programming techniques such as parallelization and high-performance computing have enabled to produce highly accurate results in these application fields. However, the runtime and hardware requirements are frequently prohibitive, such that only few simulations can be performed. Modern simulation scenarios, however, require real-time response, e.g., in simulation-based control of processes, or many-query capability, e.g., in Monte Carlo or more general statistical investigation settings. Similarly, optimization may require multiple runs of a simulation model. This is where model reduction or more general surrogate modeling comes into play by providing computational means to approximate a given expensive simulation model. This ideally enables approximate but highly accelerated simulation responses.

We focus on the approximation of vectorial functions from vectorial inputs, also known as multi-output or multi-target approximation. This is relevant in simulation models, where submodels are coupled via such vectorial interfaces. In particular, we do not modify the input dimension or output dimension of these submodels, but purely aim at surrogate modeling approaches from data sampled at these models' interfaces.

Note that this mathematical goal is different from that of projection-based model reduction, which is currently a very active field for surrogate modeling of high-dimensional systems [1, 10].

In the current article, we focus on a successful class of techniques for data-based modeling, namely kernel methods. These techniques are highly competitive or frequently superior in machine learning tasks such as classification, regression, feature extraction [9, 19, 21]. In numerical analysis, such techniques are mainly used in the context of function approximation or generalized interpolation [5, 23]. Most prominent kernels are radially symmetric; hence, these techniques comprise especially radial basis functions. Apart from empirical excellent performance, elegant analysis is possible in so-called reproducing kernel Hilbert spaces (RKHS).

We aim at function approximation based on function evaluations in given data sites. For this, the form of the approximant is a linear combination of kernel translates at certain centers. In contrast to polynomial interpolation, which in general is ill-posed in higher dimensions unless the data sites satisfy restrictive conditions (e.g., are structured in a regular grid), these kernel-based approaches allow scattered data approximation, which in the ideal case does not make any assumptions on the location of the distinct data sites. In particular, this is relevant in application cases of data-based modeling, where the data sites are given in a fixed arbitrary fashion.

In order to obtain a model that is as fast as possible, we are interested in sparsity in the linear combination, i.e., as few as possible number of nonzero terms. Sparsity in data-based modeling could be obtained, e.g., by ℓ_1-minimization, or other sparsity-inducing norms. However, here we focus on so-called greedy techniques [22], which are based on the idea of incrementally extending an approximate model until the desired accuracy is obtained. For scalar functions, greedy kernel techniques have been applied in [18]. For the case of vectorial function approximation, one should aim at a shared small set of centers among the target function components instead of constructing independent center sets for the target function dimensions. This results in overall less number of centers over the complete target vector.

We will mainly deal with the symmetric vectorial kernel orthogonal greedy algorithm (VKOGA) [24], which has proven to enable orders of magnitude of acceleration in biomechanical and multi-phase flow applications [25]. In addition to excellent performance, also convergence rates can be proven, e.g., algebraic rates [24], which are even independent of the input space dimension. In this sense, these methods are less prone to the curse of dimensionality than grid-based approaches and promise to give good approximation also in higher input dimensions. Recently, also exponential convergence for a certain greedy variant in case of infinitely smooth kernels has been proven [16].

As new methodological aspect, we present some initial results of non-symmetric greedy approximation schemes. The new method exhibits an improved stability over existing techniques, meaning that it allows larger expansion sizes in the kernel approximant, and it provides also a better accuracy in the considered experimental settings, where the input data locations are not evenly distributed in the input domain, since it allows in particular to choose the centers independently from the data sites. Although further analysis and testing are needed, these features indicate that the new method may be well suited for application in the field of surrogate modeling.

The paper is structured as follows: In Sect. 2, we recall the basic theory of scattered data interpolation and review elementary results on kernel methods. Section 3 is devoted to the presentation of greedy algorithms to produce sparse surrogate models. In particular, we recall the most relevant optimization criteria and review recent results on their approximation rate, before introducing a new non-symmetric algorithm in Sect. "A New Greedy Algorithm: Non-symmetric VKOGA". The algorithm capabilities are then demonstrated in Sect. 4, where comparisons with other kernel-based approximation techniques are also discussed. The final Sect. 5 discusses some outlook on further work.

2 Scattered Data Interpolation of High-Dimensional Functions

We represent the full model as a function $f : \Omega \subset \mathbb{R}^d \to \mathbb{R}^q$, where $d, q \in \mathbb{N}$ are the input and output dimensions, and Ω is a possibly unstructured domain of the input variables. To construct a surrogate model, we consider a set of $N \in \mathbb{N}$ data sites $X_N := \{x_1, \ldots, x_N\} \subset \Omega$ and the corresponding set of model evaluations $F_N := \{f(x_1), \ldots, f(x_N)\} \subset \mathbb{R}^q$. We refer to the pair (X_N, F_N) as *data samples*, since it represents the available information of the unknown function f. The set of input locations X_N and output values F_N are, respectively, denoted as data sample *locations* and *values*.

A general surrogate model $s_f : \Omega \to \mathbb{R}^q$ can be constructed by means of a set of continuous basis functions $\{\psi_j : \Omega \to \mathbb{R}, 1 \le j \le N\}$, by considering the ansatz

$$s_f(x)^T := \sum_{j=1}^{N} \psi_j(x)\alpha_j^T, \ x \in \Omega \tag{1}$$

for certain column vectors of coefficients $\alpha_j \in \mathbb{R}^q$. Note that we deliberately choose a transpose formulation here, as the interpolation system will naturally follow as one row per interpolation condition. To determine such coefficients, one imposes the interpolation conditions

$$s_f(x_i)^T = f(x_i)^T, \ 1 \le i \le N, \tag{2}$$

which imply, for $1 \le i \le N$,

$$s_f(x_i)^T := \sum_{j=1}^{N} \psi_j(x_i)\alpha_j^T = f(x_i)^T. \tag{3}$$

If the basis evaluations, the coefficient vectors, and the model evaluations are collected into matrices $A_\psi \in \mathbb{R}^{N \times N}$, $\alpha \in \mathbb{R}^{N \times q}$, $b \in \mathbb{R}^{N \times q}$ as follows

$$A_\psi = \begin{bmatrix} \psi_1(x_1) & \psi_2(x_1) & \dots & \psi_N(x_1) \\ \psi_1(x_2) & \psi_2(x_2) & \dots & \psi_N(x_2) \\ \vdots & \vdots & & \vdots \\ \psi_1(x_N) & \psi_2(x_N) & \dots & \psi_N(x_N) \end{bmatrix}, \ \alpha = \begin{bmatrix} \alpha_1^T \\ \alpha_2^T \\ \vdots \\ \alpha_N^T \end{bmatrix}, \ b = \begin{bmatrix} f(x_1)^T \\ f(x_2)^T \\ \vdots \\ f(x_N)^T \end{bmatrix},$$

the existence and uniqueness of a surrogate model (1) with interpolation conditions (2) is equivalent to the existence and uniqueness of a solution $\alpha \in \mathbb{R}^{N \times q}$ of the linear system

$$A_\psi \alpha = b. \tag{4}$$

To deal with concrete simulation scenarios, we need a method capable of working with possibly high-dimensional d and q, with general sets Ω and scattered data sites X_N, and still providing a well-defined and unique solution to this linear system. Kernel methods yield precisely basis functions such that the above linear system admits a unique solution for arbitrary given pairwise distinct data sites $X_N \subset \Omega$ and values F_N, for any given space dimensions d and q.

Kernel-Based Interpolation of Scalar-Valued Functions

We review some tools and basic results on kernels starting from the simple case of scalar-valued functions. What follows can be easily generalized to treat the more general multi-output case, as will be explained in "Dealing with Vector-Valued Functions" section.

Definition 1 A (strictly) positive definite kernel K on Ω is a symmetric function $K : \Omega \times \Omega \to \mathbb{R}$ such that, for all $N \in \mathbb{N}$ and for all sets of pairwise distinct points $X_N := \{x_1, \dots, x_N\} \subset \Omega$, the $N \times N$ symmetric kernel matrix

$$A_{K,X_N} := (K(x_i, x_j))_{i,j=1}^N$$

is positive semidefinite (respectively positive definite).

Notable examples of strictly positive definite kernels are generated by certain radial basis functions (RBFs), i.e., radially symmetric kernels for which there exist a function $\phi : \mathbb{R}_{\geq 0} \to \mathbb{R}$ and a positive *shape parameter* ρ such that $K(x, y) := \phi(\rho\|x - y\|_2)$ for all $x, y \in \Omega$. The most prominent examples of such functions are the Gaussian kernels, defined as $K_\rho^G(x, y) := \exp(-\rho^2\|x - y\|_2^2)$. But much more general examples exist, with different properties and in particular different smoothness. They all share the important feature to be dimension-blind, i.e., they can be evaluated by composing a cheap, one-dimensional function with the distance function, independent of the input space dimension.

We will consider here strictly positive definite kernels, since they guarantee precisely the existence of a unique solution of a surrogate model (1). Indeed, for any given set of pairwise distinct data sites $X_N \subset \Omega$, one can use the kernel to construct a data-dependent basis $\psi_j(x) := K(x, x_j), 1 \leq j \leq N$, giving the ansatz

$$s_f(x) = \sum_{j=1}^N \psi_j(x)\alpha_j := \sum_{j=1}^N K(x, x_j)\alpha_j. \tag{5}$$

Here, we can obviously omit the transpose of (3) as the α_j as well as $s_f(x)$ are scalar. The interpolation conditions (2) result in the solution of the linear system (4) (for $q = 1$) where now the system matrix is the kernel matrix A_{K,X_N}. Thanks to Definition 1, such a matrix is symmetric and positive definite and thus a unique solution always exists.

Other than providing unique solutions to the scattered data interpolation problem, kernels are directly related to certain Hilbert spaces, which turn out to be very useful to analyze the approximation process given by kernel interpolation. For an extensive treatment of this theory, we refer to the books [2, 5, 6, 23], and we recall here only the necessary tools.

Definition 2 A Hilbert space $\mathcal{H}(\Omega)$ of functions from Ω to $\mathbb{R}$ with inner product $(\cdot, \cdot)_{\mathcal{H}(\Omega)}$ is a reproducing kernel Hilbert space (RKHS) if it possesses a reproducing kernel, i.e., a symmetric function $K : \Omega \times \Omega \to \mathbb{R}$ such that

1. $K(\cdot, x) : \Omega \to \mathbb{R}$ is an element of $\mathcal{H}(\Omega)$ for all $x \in \Omega$;
2. $(f, K(\cdot, x))_{\mathcal{H}(\Omega)} = f(x)$ for all $x \in \Omega$, for all $f \in \mathcal{H}(\Omega)$.

Moreover, if a reproducing kernel exists for $\mathcal{H}(\Omega)$ it is unique and positive definite.

From a given strictly positive definite kernel K on Ω, it is always possible to construct a unique RKHS. To construct this space, one first considers the space of finite linear combinations

$$\mathcal{H}_0(\Omega) := \{f(x) := \sum_{j=1}^N K(x, x_j)\alpha_j \; : \; x_j \in \Omega, \alpha_j \in \mathbb{R}, 1 \leq j \leq N, N \in \mathbb{N}\},$$

which can be equipped with the bilinear form

$$(f, g)_K = \left(\sum_{j=1}^{N} K(x, x_j)\alpha_j, \sum_{i=1}^{M} K(x, y_i)\beta_i \right)_K := \sum_{j=1}^{N} \sum_{i=1}^{M} \alpha_j K(x_j, y_i)\beta_i,$$

which is positive definite by Definition 1. The closure of $\mathscr{H}_0(\Omega)$ with respect to the norm induced by this bilinear form is precisely the unique space where the kernel K is a reproducing kernel. It will be denoted as $\mathscr{H}_K(\Omega)$ and, when no confusion is possible, we will use the notation $(\cdot, \cdot)$ and $\| \cdot \|$, without subscript, for its inner product and norm.

It is clear from the above construction that such a space contains in particular all the functions of the form (5). Moreover, the interpolant constructed in this way is not only an element of the space $\mathscr{H}_0(\Omega)$, it also has some optimality properties with respect to $\| \cdot \|$. First, the following optimality property holds.

Proposition 1 *For a set of pairwise distinct data sites $X_N \subset \Omega$ and $f \in \mathscr{H}_K(\Omega)$, define the set of interpolants*

$$\mathscr{S} := \{s \in \mathscr{H}_K(\Omega) : s(x_i) = f(x_i), 1 \le i \le N\}.$$

Then, $\|s_f\| \le \|s\|$ for all $s \in \mathscr{S}$, i.e., the kernel interpolant (5) is the minimal norm interpolant of the given data.

Moreover, given the data sites X_N, one can form an N-dimensional subspace $V(X_N) \subset \mathscr{H}_K(\Omega)$ by considering the so-called kernel translates basis associated with X_N, i.e.,

$$V(X_N) := \mathrm{span}\{K(\cdot, x_i) : x_i \in X_N\}.$$

It follows that $s_f \in V(X_N)$. Moreover, whenever $f \in \mathscr{H}_K(\Omega)$, s_f is the orthogonal projection of f into $V(X_N)$, and in particular it is the best approximation of f with respect to $\| \cdot \|$.

Such optimality properties are particularly interesting for some specific kernels. Indeed, it can be proven that certain RBF kernels generate reproducing kernel Hilbert spaces which are norm equivalent to the Sobolev space $\mathscr{W}_2^k(\mathbb{R}^d)$. A norm minimal interpolant in $\mathscr{H}_K(\Omega)$ is thus quasi-optimal with respect to the Sobolev norm, and it can be obtained by the direct solution of a linear system, without requiring integration or evaluation of derivatives. This is particularly interesting for numerical solution of PDEs avoiding mesh-based techniques such as finite elements or finite volumes (see, e.g., [23]).

Finally, we remark that a standard way to measure the interpolation error for functions $f \in \mathscr{H}_K(\Omega)$ is provided by the following power function.

Definition 3 For a set of data sites $X_N \subset \Omega$ and a strictly positive definite kernel $K : \Omega \times \Omega \to \mathbb{R}$, the power function $P_{X_N} : \Omega \to \mathbb{R}$ is defined for $x \in \Omega$ as

$$P_{X_N}(x) := \sup_{f \in \mathcal{H}_K(\Omega), f \neq 0} \frac{|f(x) - s_f(x)|}{\|f\|},$$

where s_f is the kernel interpolant on X_N of the function $f \in \mathcal{H}_K(\Omega)$. It holds, for all $f \in \mathcal{H}_K(\Omega)$,

$$|f(x) - s_f(x)| \leq P_{X_N}(x)\|f\|, \quad x \in \Omega.$$

Moreover, $P_{X_N}(x) = 0$ if and only if $x \in X_N$.

This pointwise error bound nicely decomposes the error in a factor depending on the approximation space $V(X_N)$ but not on f, and a factor independent of $V(X_N)$ but dependent on the target function f. Furthermore, the bound is sharp.

We will see in the next sections that the power function can be computed explicitly.

Dealing with Vector-Valued Functions

The case of vector-valued functions can be obtained by extending the discussion of the previous section (see [24]).

Indeed, for $q \geq 1$ it is possible to use kernel methods to reconstruct functions $f : \Omega \to \mathbb{R}^q$ simply by considering q copies of $\mathcal{H}_K(\Omega)$, i.e., using the product space

$$\mathcal{H}_K(\Omega)^q := \{f : \Omega \to \mathbb{R}^q, (f)_j \in \mathcal{H}_K(\Omega)\}$$

equipped with the inner product

$$(f, g)_q := \sum_{j=1}^{q} (f_j, g_j).$$

The crucial point here, to have an effective method to be used in surrogate modeling, is to avoid having q different expansions, one for each component. This would result in q independent sets of centers, hence many kernel evaluations to compute a point value $s_f(x)$. To reduce the overall number of centers, one can make the further assumption that a common subspace $V(X_N)$ is used for every component. From the point of view of the actual computation of the interpolant, this is precisely equivalent to solving the linear system (4) for a coefficient matrix $\alpha \in \mathbb{R}^{N \times q}$, instead of solving q different systems with possibly q different matrices to determine each of its columns.

3 Sparse Kernel Models via Greedy Algorithms

Although kernel interpolation is always well defined as seen in the previous sections, it is often desirable to avoid solving the linear system (4). Indeed, it is well known that it can be extremely ill-conditioned for an increasing number N of points, in particular if the points in X_N are almost collapsing or if no care is taken in the selection of the shape parameter ρ when using radial basis functions. To circumvent such problems, different solution techniques have been developed in the last decade to obtain stable computations of the kernel interpolants without applying the direct solution method as it is described here (see, e.g., [2, 6–8, 11, 20]).

Moreover, the use of kernel approximants as surrogate models poses further requirements. Indeed, once the interpolant has been obtained, it is used as a surrogate of the underlying expensive model in an *online phase*, that is, a phase which requires multiple queries of the precomputed model. For a successful application in this context, the online efficiency of the methods, i.e., the computational cost required for its evaluation, is a crucial requirement. The cost of this evaluation, hence the expected speedup with respect to the full model, is essentially depending on the size of the expansion in the representation

$$s_f(x)^T = \sum_{j=1}^{N} K(x, x_j)\alpha_j^T,$$

which is equal to the number of data samples one has at hand, and can thus be very large. To avoid the evaluation of this possibly expensive sum, one can search for a reduced set of indexes $J_n \subset \{1, \ldots, N\}$, $|J_n| = n$, form the subset of data sites $X_n := \{x_i : i \in J_n\} \subset X_N$, and construct the corresponding surrogate model

$$f_n(x)^T = \sum_{j \in J_n} K(x, x_j)\alpha_j^T$$

obtained by imposing the interpolation conditions only on X_n. The index set $J_n \subset \{1, \ldots, N\}$, or equivalently the point set X_n should be such that $n \ll N$, while $\|f - f_n\|$ is ideally as small as $\|f - s_f\|$.

Greedy Methods: Symmetric VKOGA

The determination of an optimal set X_n is a combinatorial and computationally prohibitive problem which cannot be addressed in practice. A valid alternative to this global optimization problem is provided by greedy methods, which construct a nested sequence of point sets X_n by starting with the empty set $X_0 := \emptyset$, and iteratively adding one point at a time which maximizes a certain selection criterion. Observe that this process is equivalent to selecting the $n \times n$ submatrix of the $N \times N$ kernel matrix A_{K,X_N} obtained by selecting the rows and columns with indexes J_n, and computing the sparse solution obtained by the solution of the corresponding small $n \times n$ linear system.

Different selection methods are possible and they lead to different models f_n, but they all share a common structure. In particular, each method starts with a sequence of indexes $J_0 := \emptyset$, of points $X_0 := \emptyset$, and a subspace $V(X_0) := \{0\}$, and at the n-th iteration, $n \geq 1$, it selects a new index $j_n \subset \{1, \ldots, N\} \setminus J_{n-1}$ and defines

$$X_n := \{x_j : j \in J_n\} = X_{n-1} \cup \{x_{j_n}\}, \quad V(X_n) := \text{span}\{K(\cdot, x_j) : x_j \in X_n\}.$$

The sequence of interpolants f_n are obtained as solutions of the interpolation problem with data sites X_n. To avoid recomputing the complete expansion at each iteration and improve the efficiency of the method, a Newton basis is used (see [13, 14]), which is an $\mathscr{H}_K(\Omega)$-orthonormal and nested basis $\{v_j\}_{j=1}^n$ of $V(X_n)$. Using this basis, the interpolants can be written as

$$f_n(x)^T = \sum_{j=1}^n v_j(x)c_j^T = v_n(x)c_n^T + \sum_{j=1}^{n-1} v_j(x)c_j^T = v_n(x)c_n^T + f_{n-1}(x)^T,$$

i.e., only one basis element v_n and one coefficient vector c_n need to be computed at each iteration.

It is possible to prove (see [14]) that this iterative construction of the interpolant corresponds to solve the interpolation linear system using a partial pivoted Cholesky decomposition of the matrix A_{K,X_N}, where the selection rule of the indexes J_n is used as a pivoting rule. This can be done in a matrix-free way, i.e., by generating at each iteration only the single column of the full matrix corresponding to the index j_n.

This process allows also to efficiently update the pointwise residual, i.e., if we define $r_0 := f$, we have

$$r_n(x)^T := f(x)^T - f_n(x)^T = f(x)^T - f_{n-1}^T(x) - v_n(x)c_n^T = r_{n-1}(x)^T - v_n(x)c_n^T.$$

By using the Newton basis, it is also possible to efficiently update the power function (see [14]).

Proposition 2 *The power function of a sequence of data sites $\emptyset =: X_0 \subset X_1 \subset \cdots \subset X_n$ can be computed with the following iterative procedure:*

$$P_{X_0}(x)^2 = K(x, x), \quad P_{X_n}(x)^2 = K(x, x) - \sum_{j=1}^n v_j(x)^2 = P_{X_{n-1}}(x)^2 - v_n(x)^2.$$

With these tools in hand, it is easy to formulate the three main selection rules to generate the index set J_n, namely the P-, f-, and f/P-greedy selections. The three selection rules and the corresponding algorithms have been introduced, respectively, in [4] (P-greedy), [18] (f-greedy), and [12] (f/P-greedy). The formulation which is used here is the one of the VKOGA of [24], where the f- and f/P-greedy rules have been extended to the vector-valued case. We use the notation $(f(x))_j$ to denote the j-th component of a vectorial function $f : \Omega \to \mathbb{R}^q$ evaluated at $x \in \Omega$.

P-greedy: This selection rule minimizes the value of the power function by picking its current maximum. It is data-independent, meaning that only Ω and K are used, and not the function values F_N; thus, the resulting data sites X_n are good for any function $f \in \mathcal{H}_K(\Omega)$. On the other hand, it can be proven that the selected points X_n are space-filling in Ω (see [16]). The selection is defined as follows:

$$x_{j_1} := \arg \max_{x \in \Omega} \sqrt{K(x, x)}, \quad x_{j_n} := \arg \max_{x \in \Omega \setminus X_{n-1}} P_{X_{n-1}}(x), \quad n > 1.$$

f-greedy: This selection rule is data-dependent, so it selects sets of points which are good for one specific function $f \in \mathcal{H}_K(\Omega)$. The idea is to enlarge the interpolation set by adding the point where the residual attains its maximal value. Since the model is computed by interpolation, in the next iteration the residual will be zero in such a point. The selection is

$$x_{j_1} := \arg \max_{x \in \Omega} \sum_{j=1}^{q} |(f(x))_j|^2, \quad x_{j_n} := \arg \max_{x \in \Omega \setminus X_{n-1}} \sum_{j=1}^{q} |(r_{n-1}(x))_j|^2, \quad n > 1.$$

f/P-greedy: Also, this rule is data-dependent, and it is an improved version of the previous one. The selection in this case is performed by maximization of the ratio between the residual and the power function, and it can be proven that it is locally optimal, i.e., it guarantees at each step the maximal possible reduction in the interpolation error when measured with respect to $\| \cdot \|$. It can be formulated as

$$x_{j_1} := \arg \max_{x \in \Omega} \frac{\sum_{j=1}^{q} |(f(x))_j|^2}{\sqrt{K(x, x)}}, \quad x_{j_n} := \arg \max_{x \in \Omega \setminus X_{n-1}} \frac{\sum_{j=1}^{q} |(r_{n-1}(x))_j|^2}{P_{X_{n-1}}(x)}, \quad n > 1.$$

We remark that in the case when a large set of samples $X_N \subset \Omega$ is available, the above selections of points in practice are applied to X_N instead of Ω, simply by performing the maximum search over this discrete set.

Convergence Results

All the above rules can be proven to be convergent when the optimization is performed on Ω.

Theorem 1 ([4, 18, 24]) *Let $f \in \mathcal{H}_K(\Omega)$, $\{X_n\}_n \subset \Omega$ be the sequence of data sites produced either by the P-, f-, or f/P-greedy algorithm and $\{f_n\}_n$ the corresponding sequence of interpolants. Then,*

$$\lim_{n \to \infty} \|f - f_n\|_{L_\infty(\Omega)} = 0.$$

Convergence rates are also available.

The P-greedy algorithm can be proven to achieve the same convergence order of the best-known kernel approximation scheme in the case of RBF kernels. Such orders depend on the asymptotic decay of the Fourier transform of the kernels (see [17]), which is in turn related to the smoothness of ϕ. Here, we just refer to these orders as exponential and algebraic, and we refer to the paper [16] for further details. Nevertheless, we recall that the algebraic decay rate applies, for example, to the finitely smooth compactly supported Wendland kernels (and it is known to be quasi-optimal), while the exponential one applies to the infinitely smooth Gaussian or multi-quadric kernels.

Theorem 2 ([16]) *If Ω is bounded and satisfies an interior cone condition and $K(x, y) := \phi(\|x - y\|)$ is an RBF kernel, the VKOGA P-greedy algorithm computes point sets $X_n \subseteq \Omega$ with the following decay of the power function.*

1. *If $\phi \in \mathscr{C}^\beta(\Omega)$, $\beta > d/2$, then the P-greedy points satisfy*

$$\|P_{X_n}\|_{L_\infty(\Omega)} \le c_1 n^{-\frac{\beta}{d}+\frac{1}{2}}.$$

2. *If ϕ is infinitely smooth in Ω, then the P-greedy points satisfy*

$$\|P_{X_n}\|_{L_\infty(\Omega)} \le c_2 \exp(-c_3 n^{1/d}).$$

The constants c_1, c_2, c_3 do not depend on n.

In the case of the f-greedy algorithm, only a convergence of order $1/d$ is known (see [12]) and it applies so far only to scalar-valued functions, but the following extension to the vectorial case is straightforward.

Theorem 3 ([12]) *Let Ω be bounded and satisfy an interior cone condition, and let K be an RBF kernel with $\phi \in C^2(\Omega)$. The points produced by the f-greedy algorithm provide an interpolation error such that, for all $1 \le j \le q$,*

$$\min_{1\le i\le n} \|(f - f_i)_j\|_{L_\infty(\Omega)} \le cn^{-1/d}.$$

The constant c depends on q but not on n or j.

The known convergence rate for the f/P-greedy algorithm instead does not require restrictions on the kernel, but it applies to a specific function class. The convergence rate is slower than the one for the P-greedy algorithm, but it has the advantage of being independent of the input space dimension d. Therefore, it raises the hope to be less prone to the curse of dimensionality. For simplicity, we state the following theorem in the case $K(x, x) = 1$ for all $x \in \Omega$, which comes for free in the case of RBF kernels, and point to [24] for further details.

Theorem 4 ([24]) *For $M > 0$, let*

$$\mathcal{H}_K(\Omega)_M^q := \left\{ f \in \mathcal{H}_K(\Omega)^q : (f)_j := \sum_{i=1}^N K(x, x_i)(\alpha_i)_j, \sum_{i=1}^\infty |(\alpha_i)_j| \le M, 1 \le j \le q \right\}$$

be the subset of $\mathcal{H}_K(\Omega)^q$ of functions whose coefficient sequences $\{(\alpha_i)_j\}_{i=1}^\infty$ have ℓ_1 norm bounded by M for all $j = 1, \ldots, q$. For any $f \in \mathcal{H}_K(\Omega)_M^q$, the f/P-greedy approximant sequence $\{f_n\}_n$ gives an error

$$\|f - f_n\| \le \sqrt{q} M \left(1 + \frac{n}{q}\right)^{-\frac{1}{2}}. \tag{6}$$

Note that, despite the dimension independence of the bound (6), the set $\mathcal{H}_K(\Omega)_M^q$ is dependent on the dimension via the possible coefficient vectors.

We remark that the above convergence results seem to be far from being exhaustive for the f- and f/P-greedy algorithms. Indeed, while the theoretical convergence of the P-greedy algorithm in Theorem 2 matches the observed numerical results, in the last two cases the numerically observed convergence rates of the methods are much faster than the one reported in Theorems 3 and 4.

A New Greedy Algorithm: Non-symmetric VKOGA

We briefly present here a new greedy algorithm which extends the above strategies. The following discussion is a preliminary investigation, which will be further developed in an upcoming paper [15].

The motivation for this extension comes from the fact that the above methods construct an interpolant which is defined by basis functions $\{K(\cdot, x_j)\}_{j \in J_n}$ with points x_j which are selected in practice from the sampling points X_N, and not from the continuous set Ω. In the case X_N is not dense enough in Ω, as is often expected in applications, the resulting model will be much different from the one obtained by performing the greedy selection on the full set, and thus, the convergence to the exact model can be spoiled. For this reason, we aim at a different formulation of the interpolation problem, namely we consider an interpolant of the form

$$s_f(x)^T = \sum_{j=1}^M K(x, y_j)\alpha_j^T,$$

where now the kernel basis $K(x, y_j)$ is defined using a different point set $Y_M \subset \Omega$ as centers, possibly with $M \ne N$ and different from the set of data sample sites X_N. Also in this case, the interpolation conditions are defined as

$$s_f(x_i)^T = f(x_i)^T, \quad 1 \le i \le N,$$

and we do not need to know the values of f on the points Y_M, which can thus be chosen freely.

On the other hand, the resulting linear system has the matrix

$$
A_{K,X_N,Y_N} = \begin{bmatrix}
K(x_1, y_1) & K(x_1, y_2) & \dots & K(x_1, y_M) \\
K(x_2, y_1) & K(x_2, y_2) & \dots & K(x_2, y_M) \\
\vdots & \vdots & & \vdots \\
K(x_N, y_1) & K(x_N, y_2) & \dots & K(x_N, y_M)
\end{bmatrix},
$$

which is no more guaranteed to be positive definite (indeed, it is no longer symmetric, nor square if $M \neq N$), and it can also be singular for bad choices of Y_M. Nevertheless, a greedy selection strategy can be designed also in this setting to construct a uniquely defined interpolant.

We first introduce the analytical tools to deal with this interpolant, and we will come back later to a specific selection strategy. We assume to have a set of data sites $X_N \subset \Omega$ on which we know the evaluations F_N of an otherwise unknown function $f \in \mathscr{H}_K(\Omega)^q$. Furthermore, we assume to have a set $Y_M \subset \Omega$ of possible centers for the kernel basis. Observe that we can consider sets with $M \gg N$, and no restriction is needed since they are not connected to the function samples we have. From those sets, for $n \geq 0$ we select two sequences of nested point sets $\emptyset =: X_0 \subset X_1 \subset X_n \subset \dots X_N$ and $\emptyset =: Y_0 \subset Y_1 \subset Y_n \subset \dots Y_M$, represented by two index sets $I_n := \{i_1, \dots, i_n\} \subset \{1, \dots, N\}$ and $J_n := \{j_1, \dots, j_n\} \subset \{1, \dots, M\}$. We then define the sequence of interpolants $\{f_n\}_n$ through the basis $\{K(x, y_j) : j \in J_n\}$ defined by Y_n, i.e.,

$$
f_n(x)^T := \sum_{j \in J_n} K(x, y_j) \alpha_j^T
$$

with coefficients determined by the following interpolation conditions on X_n

$$
f_n(x_i)^T = f(x_i)^T, \ i \in I_n.
$$

These two requirements imply that the coefficients are, as in the other cases, the solution of a linear system, but now involving a non-symmetric kernel matrix A_{K,X_n,Y_n}, which is in turn the $n \times n$ submatrix of the matrix A_{K,X_N,Y_N} defined by row indexes I_n and column indexes J_n.

On the Hilbert space side, point selection means to construct two different sequences of subspaces $V(X_n) := \mathrm{span}\{K(\cdot, x_i) : i \in I_n\}$ and $V(Y_n) := \mathrm{span}\{K(\cdot, y_j) : j \in J_n\}$. We first see how to extend the Newton basis to this case.

Definition 4 Let X_n, Y_n, $V(X_n)$, $V(Y_n)$ be defined as above, assume A_{K,X_n,Y_n} is invertible, and let $A_{K,X_n,Y_n} = LU$ be an LU decomposition with $L_{ii} = 1$ (i.e., a partial LU decomposition of A_{K,X_N,Y_M} obtained with pivoting rule corresponding to row indexes I_n and column indexes J_n).

The Newton bases $\{v_i\}_{i=1}^n$ of $V(X_n)$ and $\{u_j\}_{j=1}^n$ of $V(Y_n)$ are defined as

$$v_i = \sum_{k=1}^n \beta_{ki} K(\cdot, x_{i_k}), \quad u_j = \sum_{l=1}^n \gamma_{lj} K(\cdot, y_{j_l})$$

where $(L^{-T})_{ki} = \beta_{ki}$, $(U^{-1})_{lj} = \gamma_{lj}$, $1 \le k, i, l, j \le n$.

These Newton bases are indeed bases of the two subspaces and satisfy a bi-orthogonality property. The assumption $L_{ii} = 1$ is not used here.

Proposition 3 *The Newton bases of Definition 4 are bases of $V(X_n)$, $V(Y_n)$, and they satisfy*

$$(v_i, u_j) = \delta_{ij}.$$

Proof For notational simplicity, we assume in the following that $I_n = J_n = \{1, \ldots, n\}$. From the above definition, we have

$$v_i = \sum_{k=1}^n \beta_{ki} K(\cdot, x_k) = \sum_{k=1}^i \beta_{ki} K(\cdot, x_k), \quad u_j = \sum_{l=1}^n \gamma_{lj} K(\cdot, y_l) = \sum_{l=1}^j \gamma_{lj} K(\cdot, y_l)$$

for the coefficients of Definition 4, i.e., $(L^{-T})_{ki} = \beta_{ki}$ and $(U^{-1})_{lj} = \gamma_{lj}$, $1 \le k, i, l, j \le n$.

Since U is invertible and $\{K(\cdot, x_i)\}_{i=1}^n$ is a basis of $V(X_n)$, also $\{v_i\}_{i=1}^n$ is a basis of $V(X_n)$, and the same holds for $\{u_j\}_{j=1}^n$ and $V(Y_n)$ thanks to the invertibility of L. Moreover, for all $1 \le i, j \le n$ we can use the definition of the inner product of $\mathscr{H}_K(\Omega)$ to compute

$$(v_i, u_j) = \left(\sum_{k=1}^n \beta_{ki} K(\cdot, x_k), \sum_{l=1}^n \gamma_{lj} K(\cdot, y_l) \right) = \sum_{k,l=1}^n \beta_{ki} \gamma_{lj} \left(K(\cdot, x_k), K(\cdot, y_l) \right)$$

$$= \sum_{k,l=1}^n \beta_{ki} \gamma_{lj} K(x_k, y_l) = (L^{-1} A_{K, X_n, Y_n} U^{-1})_{ij} = (L^{-1} L U U^{-1})_{ij} = \delta_{ij},$$

and we obtain bi-orthogonality. $\square$

We remark that, since the matrices of change of basis are lower triangular, the j-th element of both bases depends only on the first j points in X_n and Y_n, so the basis can be incrementally updated.

With the extended Newton bases, the non-symmetric interpolants can be computed in the following way.

Proposition 4 *Assume $X_{n-1} \subset X_N$, $Y_{n-1} \subset Y_M$ are already computed and assume the interpolation problem with data sites X_{n-1} and kernel centers Y_{n-1} is well defined.*

For any choice of $x_{i_n} \in X_N \setminus X_{n-1}$, $y_{j_n} \in Y_M \setminus Y_{n-1}$, the interpolation problem with data sites $X_n := X_{n-1} \cup \{x_{i_n}\}$ and kernel centers $Y_n := Y_{n-1} \cup \{y_{j_n}\}$ is well defined if and only if

$$v_n(y_{j_n}) \neq 0. \tag{7}$$

In this case, the interpolant f_n can be computed as

$$(f_n(x))_k = \sum_{j=1}^{n} ((f)_k, v_j) u_j(x) = (f_{n-1}(x))_k + ((f)_k, v_n) u_n(x), \quad 1 \leq k \leq q. \tag{8}$$

Proof We assume again, for simplicity, that $I_n = \{1, \ldots, n\} = J_n$. The interpolation problem has a unique solution if and only if the matrix A_{K,X_n,Y_n} is invertible, and this condition is equivalent to having an LU decomposition with invertible factors. Since we assumed $L_{ii} = 1$, this means to check that $U_{ii} \neq 0$, $1 \leq i \leq n$. From Definition 4, for every basis element v_i, $1 \leq i \leq n$ and point y_j, $1 \leq j \leq n$, we have

$$v_i(y_j) = \sum_{k=1}^{n} \beta_{ki} K(y_j, x_k) = (L^{-T} A_{K,X_n,Y_n})_{ij},$$

so $U_{ij} = v_i(y_j)$, and thus it suffices to check condition (7) to guarantee the existence of a unique solution.

We now assume that the interpolant with X_n, Y_n is well defined and prove that formula (8) is a valid representation for it. We restrict to the scalar case $q = 1$, since the general case follows by applying the same argument to each component $(f)_k$, $1 \leq k \leq q$. We use the Definition 4 to obtain

$$f_n(x) = \sum_{j=1}^{n} (f, v_j) u_j(x) = \sum_{j=1}^{n} \left(f, \sum_{k=1}^{n} \beta_{kj} K(\cdot, x_k) \right) \sum_{l=1}^{n} \gamma_{lj} K(x, y_l) \tag{9}$$

$$= \sum_{j=1}^{n} \sum_{k=1}^{n} \beta_{kj} (f, K(\cdot, x_k)) \sum_{l=1}^{n} \gamma_{lj} K(x, y_l) = \sum_{j=1}^{n} \sum_{k=1}^{n} \beta_{kj} f(x_k) \sum_{l=1}^{n} \gamma_{lj} K(x, y_l)$$

$$= \sum_{k=1}^{n} f(x_k) \sum_{j,l=1}^{n} \beta_{kj} \gamma_{lj} K(x, y_l).$$

First observe that the following relation holds for each $1 \leq i \leq n$:

$$\sum_{j,l=1}^{n} \beta_{kj} \gamma_{lj} K(x_i, y_l) = \sum_{j,l=1}^{n} (L^{-T})_{kj} (U^{-1})_{lj} A_{il} = \sum_{j=1}^{n} (L^{-T})_{kj} \sum_{l=1}^{n} A_{il} (U^{-1})_{lj}$$

$$= \sum_{j=1}^{n} (L^{-T})_{kj} (AU^{-1})_{ij} = \sum_{j=1}^{n} (L^{-1})_{jk} (AU^{-1})_{ij} = (AU^{-1}L^{-1})_{ik} = \delta_{ik}.$$

Using this equality and the expression (9), we can evaluate the interpolation conditions $f_n(x_i) = f(x_i)$, $1 \leq i \leq n$, and we obtain indeed

$$f_n(x_i) = \sum_{k=1}^{n} f(x_k) \sum_{j,l=1}^{n} \beta_{kj} \gamma_{lj} K(x_i, y_l) = \sum_{k=1}^{n} f(x_k) \delta_{ik} = f(x_i),$$

and thus, the expression (8) defines the unique interpolant.   □

This formulation of the interpolant motivates our new greedy selection criterion. We can select the sample point x_{i_n} by residual maximization as in the f-greedy algorithm, i.e.,

$$x_{i_n} := \arg\max_{x \in \Omega \setminus X_{n-1}} \sum_{j=1}^{q} |(r_{n-1}(x))_j|^2.$$

Then, the kernel center y_{j_n} can be selected to maximize the condition (7), i.e.,

$$y_{j_n} := \arg\max_{y \in \Omega \setminus Y_{n-1}} |v_n(y)|.$$

4 Experiments

We now test the symmetric and non-symmetric greedy methods from different points of view. We will first investigate their application on artificial data in order to test their features in an ideal environment. Then, we will construct surrogate models in a real simulation scenario.

Example of Application on Artificial Data

The two following experiments aim at comparing the f-greedy and the non-symmetric greedy algorithm on an artificial problem.

We use the Gaussian kernel with fixed shape parameter $\rho := 1$ in the unit square $\Omega = [0, 1]^2$. The scalar target function $f : \Omega \to \mathbb{R}$ is defined as the linear combination of $m = 100$ translates of the same kernel with respect to a set $Z \subset \Omega$ of random points and with random coefficients, i.e.,

$$f(x) := \sum_{i=1}^{m} K(x, z_i)\alpha_i, \quad Z := \{z_1, \ldots, z_m\}.$$

By this, the error can later be measured in the RKHS norm $\| \cdot \|$. The functions used in the two following examples are illustrated on the right in Figs. 1 and 4.

Both symmetric and non-symmetric algorithms are stopped by imposing a maximal expansion size of $n_{max} = 300$.

In both experiments, we consider N random sampling sites X_N for the function evaluation and a set of $\bar{M}$ equally spaced points $\bar{Y}$ to select the centers in the non-symmetric method. The set Y_M is then modified to include also the sample sites X_N; i.e., $Y_M := X_N \cup \bar{Y}$ is used.

In the first case, we consider a set of samples X_N which is well distributed in Ω (see Fig. 1, left). The two algorithms select different sets of points, which are plotted in Fig. 2, and the resulting errors as functions of the size n of the surrogate models are

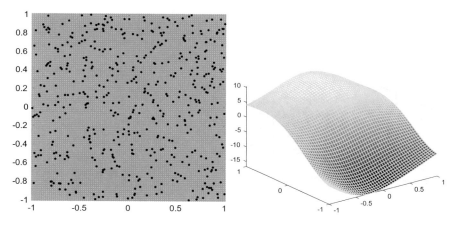

Fig. 1 Left: sampling points X_N (black) and candidate centers $\bar{Y}$ (gray). Right: target function f

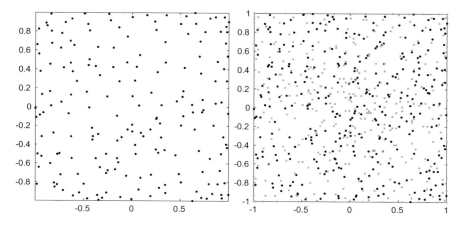

Fig. 2 Points selected by the greedy algorithms. Left: X_n for symmetric method. Right: points for non-symmetric method (black: sampling points X_n; gray: centers Y_n)

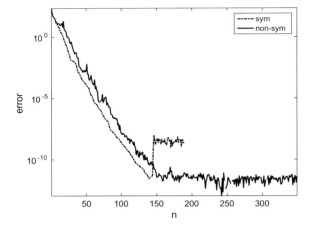

Fig. 3 Decay of the error $\| f - f_n \|$ in the RKHS norm w.r.t the expansion size n. Symmetric method (dotted line) and the non-symmetric method (solid line)

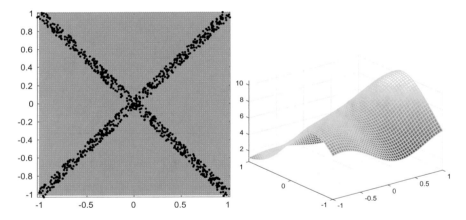

Fig. 4 Left: sampling points X_N (black) and candidate centers $\bar{Y}$ (gray). Right: target function f

shown in Fig. 3. Note that we can evaluate the exact RKHS error without involving a test set, as we know $f \in \mathcal{H}_K(\Omega)$. In this case, the two methods behave similarly, but the non-symmetric one is able to construct a larger expansion without instability problems, which instead occurs for the symmetric method.

In the second case, instead, the function sample locations X_N are limited to a cross-shaped region of Ω (see Fig. 4, left). In this case, the symmetric method is not able to select points outside this cross, and thus, we should expect a worse error in approximating a function which is defined on the whole domain Ω. This indeed can be observed in the selected points (see Fig. 5). Indeed, the symmetric method selects centers and sample sites to be identical, and they are forced to be on the cross-like

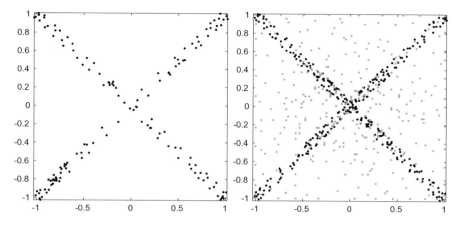

Fig. 5 Points selected by the greedy algorithms. Left: X_n for symmetric method. Right: points for non-symmetric method (black: sampling points X_n; gray: centers Y_n)

Fig. 6 Decay of the error $\|f - f_n\|$ in the RKHS norm w.r.t the expansion size n. Symmetric method (dotted line) and the non-symmetric method (solid line)

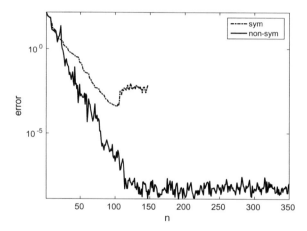

domain. The non-symmetric one instead can select centers from the whole Ω, and it does so, by covering it. The behavior is well reflected also in the error decay (see Fig. 6), where the non-symmetric method has a faster decay rate.

Note that in the decay phase, the convergence of both methods is exponential. This is similar to the provable convergence rate of the P-greedy VKOGA [16]. Thus, we expect that also for the symmetric f-greedy and non-symmetric VKOGA procedures, exponential convergence holds.

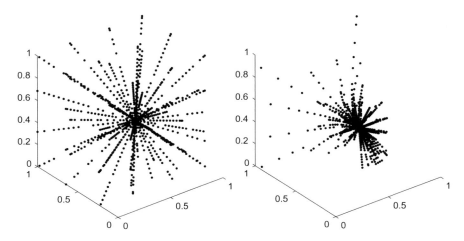

Fig. 7 Normalized training for human spine simulation. Input data X_N (left) and output data F_N (right)

Human Spine Simulation

We move now to a real simulation scenario of modeling of the human spine. The use of greedy kernel algorithms to produce a surrogate model of this simulation has been extensively studied in [25], to which we refer for further details. The full model is a function $f : \mathbb{R}^3 \to \mathbb{R}^3$ which represents the coupling of a multi-body system (MBS) simulator with a FEM simulation representing an intervertebral disk. In particular, the inputs of f are geometric parameters, while the target values of f are physical forces/moments to be used in the MBS simulation. We compare here the surrogate models obtained by the f-greedy, the non-symmetric greedy algorithm, and by support vector regression (SVR) (see, e.g., [19]).

The dataset $\mathscr{D} = (X_N, F_N)$ is produced by running the full model simulation for a set of $N = 1433$ data sites $X_N \subset \mathbb{R}^3$, giving outputs $F_N \subset \mathbb{R}^{N \times 3}$ (Fig. 7 and Table 1). We report in this table also the expected execution time of the full model, which has been estimated in [25]. We remark that this time has been obtained by estimating the number of queries to the function f required to run the full time-dependent MBS simulation. It is estimated that this simulation requires 720000 queries of f, since the full model contains 24 disks and the system is simulated for $30s$ with a timestep of $10^{-3}s$. In [25], a speedup of about 180000 was reported for a similar kernel model.

The dataset is randomly permuted and divided into training set $\mathscr{D}_{tr}$, validation set $\mathscr{D}_{val}$, and test set $\mathscr{D}_{te}$, containing, respectively, about the 80%, 10%, 10% of the total number of data.

The symmetric and non-symmetric greedy methods are executed with the Gaussian kernel. The shape parameter is chosen by validation on $\mathscr{D}_{val}$ of the model trained on $\mathscr{D}_{tr}$ by trying $n_\rho := 100$ logarithmically equally spaced values in

Table 1 Dataset for the human spine simulation

N	Bounds for X_N	Bounds for F_N	$|\mathscr{D}_{tr}|$	$|\mathscr{D}_{val}|$	$|\mathscr{D}_{te}|$	Extrapolated runtime (720000 runs)
1433	$[-2.5, 2.5] \times [-2.7, 3.5] \times [-6, 6]$	$[-0.2, 0.2] \cdot 10^4 \times [-0.15, 0.25] \cdot 10^4 \times [-1.8, 1.8] \cdot 10^4$	1147	143	143	Estimated 600 h

$[\rho_{\min} := 10^{-4}, \rho_{\max} := 2]$. For both methods, the two optimal parameters ρ^* giving the smallest error on the validation set are then used to train the models on $\mathscr{D}_{tr} \cup \mathscr{D}_{val}$.

The error is then measured on $\mathscr{D}_{te}$ as

$$\text{mean}_{(x_i, f_i) \in \mathscr{D}_{te}} \| f_n(x_i) - f_i \|_2^2,$$

i.e., we consider the mean over the individual samples of the squared $\| \cdot \|_2$-error on $\mathbb{R}^q$. The same error is used for the validation step for all the algorithms.

Both symmetric and non-symmetric algorithms are stopped by imposing a maximal expansion size of $n_{max} := 500$, a tolerance $\tau_f := 10^{-12}$ on the residual, and a stability tolerance $\tau_s := 10^{-14}$ on the power function (symmetric) and on $|v_j(y_j)|$ (non-symmetric).

The non-symmetric algorithm uses a set $\bar{Y}$ of $\bar{M}$ equally spaced points in the hypercube $[-2, 3]^3$ enclosing X_N. The set Y_M is then modified to include also the sampling sites, i.e., $Y_M := X_N \cup \bar{Y}$ is used, resulting in $M = 11795$ points.

As an SVR implementation, we use *libsvm*, Version 3.22 [3] and we use here ε-SVR with the Gaussian kernel. The algorithm depends on a regularization parameter C, on the kernel width ρ and on ε. The first one is fixed to the default value $C := 1$. The two parameters ρ and ε are validated using the same validation set as before and with the same error measure. The validation is performed on $n_\rho := n_\varepsilon := 30$ logarithmically equally spaced values in $[\rho_{\min} := 10^{-4}, \rho_{\max} := 2]$ and $[\varepsilon_{\min} := 10^{-4}, \varepsilon_{\max} := 10]$, respectively. Furthermore, we used a tolerance $\tau := 10^{-12}$ for the termination criterion. Observe that in this case, we trained three different models, one for each output of the full model f, and the three models are then combined to obtain a unique model $f_n : \mathbb{R}^3 \to \mathbb{R}^3$. We use this model to select the optimal parameters ρ^* and ε^* giving the smallest error on the validation set, which are then used to train the models on $\mathscr{D}_{tr} \cup \mathscr{D}_{val}$. Note that the expansion size n cannot be set a priori but the number of support vectors is an output of the training step depending on the choice of the parameters.

For all methods, the training and validation sets $\mathscr{D}_{tr} = (X_{N,tr}, F_{N,tr})$ and $\mathscr{D}_{val} = (X_{N,val}, F_{N,val})$ are jointly scaled such that $X_{N,tr} \cup X_{N,val} \subset [0, 1]^d$ and $F_{N,tr} \cup F_{N,val} \subset [0, 1]^q$. Instead, $\mathscr{D}_{te}$ is not scaled, and the prediction is made by scaling the input of each surrogate model and its output, according to the scaling computed for the train and validation sets.

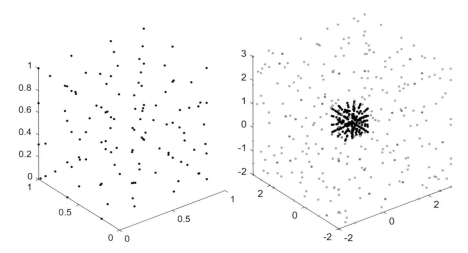

Fig. 8 Points selected by the greedy algorithms. Left: X_n for symmetric method. Right: points for non-symmetric method (black: sampling points X_n; gray: centers Y_n)

The points selected by the two greedy methods are illustrated in Fig. 8, and it is clear that the non-symmetric method is able to select points which cover the full space enclosing the training set.

The resulting optimal parameters, the size of the expansions, the errors computed on the test set, and the evaluation times (which is computed as the average time over 10^4 runs with random input values) of the models are reported in Table 2. The times are already multiplied by a factor 720000, to have a comparison with the full model execution time. Still note that the absolute runtime cannot be directly compared between the two tables due to different computer platforms. The symmetric method S-VKOGA reaches an expansion size of $n = 116$, but then the test error is increasing (see Fig. 9), so we report the values for $n = 100$. It is worth pointing out that the expansion sizes, thus the evaluation times, of the two greedy methods are quite smaller than the ones of the SVR approximant, which furthermore selects different support vectors for the three output dimensions. Moreover, the error is of about 2 order of magnitude smaller. In any case, the output is of magnitude 10^4, so also the results of the SVR model are to be considered good.

From this test, the non-symmetric method NS-VKOGA demonstrates to be an improvement over the symmetric one, and both of them outperform ε-SVR, producing considerably smaller (i.e., cheaper to evaluate) and more accurate models.

Fig. 9 Decay of the test error w.r.t. the expansion size n for the symmetric method (dotted line) and non-symmetric method (solid line)

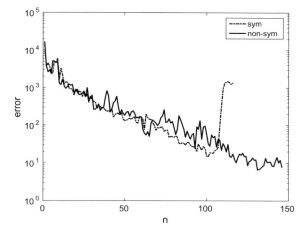

Table 2 Expansion sizes, optimal parameters, errors, and runtimes obtained with the three algorithms

Method	ρ^*	ε^*	n	Error	Extrapolated runtime (720000 runs) (s)
S-VKOGA	0.37	–	100	$1.54e + 01$	1.67
NS-VKOGA	0.22	–	146	$7.57e + 00$	1.98
ε-SVR	0.13	4.52	311	$2.51e + 03$	21.1
			614		
			605		

5 Outlook

In this paper, we discussed and wanted to promote the use of kernel methods, and especially of greedy algorithms, to produce accurate and cheap to evaluate surrogate models of high-dimensional models (in input and output). We presented some basic tools and reviewed some recent developments on the topic, which demonstrate a solid theoretical foundation of this approach and in particular a quasi-optimal approximation rate for certain selection rules. We also presented initial results of a new non-symmetric greedy method, which experimentally shows promising features.

Several questions remain open. First, the observed convergence rate of the f- and f/P-greedy algorithms is much better than the currently provable error bounds presented in Sect. 3. Their data-dependent nature makes them especially attractive to recover functions with high-dimensional inputs, where a smaller expansion size can be expected than in the case of the space-filling P-greedy algorithm.

Moreover, the analysis of the non-symmetric greedy algorithm is still preliminary, although the numerical experiments are promising and suggest that this method improves the symmetric ones. Further investigation is needed on this side.

Acknowledgements Both authors would like to thank the German Research Foundation (DFG) for financial support within the Cluster of Excellence in Simulation Technology (EXC 310) at the University of Stuttgart.

References

1. Antoulas, A.: Approximation of Large-Scale Dynamical Systems. SIAM Publications, Philadelphia, PA (2005)
2. Buhmann, M.D., Dinew, S., Larsson, E.: A note on radial basis function interpolant limits. IMA J. Numer. Anal. **30**(2), 543–554 (2010)
3. Chang, C.-C., Lin, C.-J.: LIBSVM: A Library For Support Vector Machines. Software verfgbar unter. http://www.csie.ntu.edu.tw/~cjlin/libsvm (2001)
4. De Marchi, S., Schaback, R., Wendland, H.: Near-optimal data-independent point locations for radial basis function interpolation. Adv. Comput. Math. **23**(3), 317–330 (2005)
5. Fasshauer, G. E.: Meshfree Approximation Methods with MATLAB, volume 6 of Interdisciplinary Mathematical Sciences. With 1 CD-ROM. Windows, Macintosh and UNIX. World Scientific Publishing Co., Pte. Ltd., Hackensack, NJ (2007)
6. Fasshauer, G.E., McCourt, M.J.: Stable evaluation of Gaussian radial basis function interpolants. SIAM J. Sci. Comput. **34**(2), A737–A762 (2012)
7. Fornberg, B., Larsson, E., Flyer, N.: Stable computations with Gaussian radial basis functions. SIAM J. Sci. Comput. **33**(2), 869–892 (2011)
8. Fornberg, B., Wright, G., Larsson, E.: Some observations regarding interpolants in the limit of flat radial basis functions. Comput. Math. Appl. **47**(1), 37–55 (2004)
9. Haasdonk, B.: Transformation knowledge in pattern analysis with kernel methods, distance and integration kernels. Ph.D. thesis, Albert-Ludwigs-Universität, Freiburg im Breisgau, Fakultät für Angewandte Wissenschaften, Mai. Published 2006 as ISBN-3-8322-5026-3, Shaker-Verlag, Aachen, and Online at http://www.freidok.uni-freiburg.de/volltexte/2376 (2005)
10. Haasdonk, B.: Reduced basis methods for parametrized PDEs—a tutorial introduction for stationary and instationary problems. In: Benner, M.O.P., Cohen, A., Willcox, K. (eds.) Model Reduction and Approximation: Theory and Algorithms. SIAM, Philadelphia (2017)
11. Larsson, E., Fornberg, B.: Theoretical and computational aspects of multivariate interpolation with increasingly flat radial basis functions. Comput. Math. Appl. **49**(1), 103–130 (2005)
12. Müller, S.: Komplexität und Stabilität von kernbasierten Rekonstruktionsmethoden. Ph.D. thesis, Fakultät für Mathematik und Informatik, Georg-August-Universität Göttingen (2009)
13. Müller, S., Schaback, R.: A Newton basis for kernel spaces. J. Approx. Theory **161**(2), 645–655 (2009)
14. Pazouki, M., Schaback, R.: Bases for kernel-based spaces. J. Comput. Appl. Math. **236**(4), 575–588 (2011)
15. Santin, G., Haasdonk, B.: Non-symmetric kernel greedy interpolation. University of Stuttgart, in preparation (2017)
16. Santin, G., Haasdonk, B.: Convergence rate of the data-independent P-greedy algorithm in kernel-based approximation. Dolomit. Res. Notes Approx. **10**, 68–78 (2017)
17. Schaback, R.: Error estimates and condition numbers for radial basis function interpolation. Adv. Comput. Math. **3**(3), 251–264 (1995)
18. Schaback, R., Wendland, H.: Numerical techniques based on radial basis functions. In: Curve and Surface Fitting: Saint-Malo 1999, Vanderbilt University Press, pp. 359–374 (2000)
19. Schölkopf, B., Smola, A.: Learning with Kernels. The MIT Press (2002)
20. Song, G., Riddle, J., Fasshauer, G.E., Hickernell, F.J.: Multivariate interpolation with increasingly flat radial basis functions of finite smoothness. Adv. Comput. Math. **36**(3), 485–501 (2012)

21. Steinwart, I., Christmann, A.: Support Vector Machines, Information Science and Statistics. Springer, New York (2008)
22. Temlyakov, V.N.: Greedy approximation. Acta Numer. **17**, 235–409 (2008)
23. Wendland, H.: Scattered Data Approximation. Cambridge Monographs on Applied and Computational Mathematics, vol. 17. Cambridge University Press, Cambridge (2005)
24. Wirtz, D., Haasdonk, B.: A vectorial kernel orthogonal greedy algorithm. Dolomites Res. Notes Approx. **6**:83–100 (2013). (Proceedings of DWCAA12)
25. Wirtz, D., Karajan, N., Haasdonk, B.: Surrogate modelling of multiscale models using kernel methods. Int. J. Numer. Methods Eng. **101**(1), 1–28 (2015)

Set-Oriented Multiobjective Optimal Control of PDEs Using Proper Orthogonal Decomposition

Dennis Beermann, Michael Dellnitz, Sebastian Peitz and Stefan Volkwein

Abstract In this chapter, we combine a global, derivative-free subdivision algorithm for multiobjective optimization problems with a posteriori error estimates for reduced-order models based on Proper Orthogonal Decomposition in order to efficiently solve multiobjective optimization problems governed by partial differential equations. An error bound for a semilinear heat equation is developed in such a way that the errors in the conflicting objectives can be estimated individually. The resulting algorithm constructs a library of locally valid reduced-order models online using a Greedy (worst-first) search. Using this approach, the number of evaluations of the full-order model can be reduced by a factor of more than 1000.

1 Introduction

Many problems in engineering and physics can be modeled by partial differential equations (PDEs), starting from fairly simple problems such as the linear heat equation up to highly nonlinear fluid flow problems governed by the Navier–Stokes equations. When designing an application where the underlying dynamical system is given by a PDE, we are faced with a PDE-constrained optimal control problem [28]. Due to the ever-increasing complexity of technical systems and design requirements, there are nowadays only few problems, where only one objective is of importance. For example, in buildings we want to provide a comfortable room temperature while at the same time minimizing the energy consumption. This example illustrates that

D. Beermann · S. Volkwein (✉)
Department of Mathematics and Statistics, University of Konstanz, Konstanz, Germany
e-mail: stefan.volkwein@uni-konstanz.de

D. Beermann
e-mail: dennis.beermann@uni-konstanz.de

M. Dellnitz · S. Peitz
Department of Mathematics, Paderborn University, Paderborn, Germany
e-mail: dellnitz@math.upb.de

S. Peitz
e-mail: speitz@math.upb.de

© Springer International Publishing AG, part of Springer Nature 2018
W. Keiper et al. (eds.), *Reduced-Order Modeling (ROM) for Simulation and Optimization*,
https://doi.org/10.1007/978-3-319-75319-5_3

many objectives are often equally important and also contradictory such that we are forced to accept a trade-off between them. This results in a *multiobjective optimization problem* (MOP), where multiple objectives have to be minimized at the same time. Similar to scalar optimization problems, we want to find an optimal solution to this problem. However, in a multiobjective optimization problem, we have to identify the set of *optimal compromises*, the so-called *Pareto set*.

Multiobjective optimization is an active area of research. Different approaches exist to address MOPs, e.g., *deterministic approaches* [10, 18], where ideas from scalar optimization theory are extended to the multiobjective situation. In many cases, the resulting numerical method involves the consecutive solution of multiple scalar optimization problems. *Continuation methods* make use of the fact that under certain smoothness assumptions the Pareto set is a manifold [14]. Another prominent approach is based on *evolutionary algorithms* [6], where the underlying idea is to evolve an entire set of solutions (population) during the optimization process. *Set-oriented methods* provide an alternative deterministic approach to the solution of MOPs. Utilizing subdivision techniques (cf. [7, 26]), the desired Pareto set is approximated by a nested sequence of increasingly refined box coverings.

When addressing PDE-constrained MOPs, many evaluations of this PDE are required and hence, the computational effort quickly becomes prohibitively large. The typical procedure is to discretize the spatial domain by a numerical mesh, which transforms the infinite-dimensional into a (potentially very large) finite-dimensional system (i.e., a system of coupled ordinary differential equations). With increasing computational capacities, the size of problems that can be solved has increased tremendously during the last decades [24]. However, many technical applications result in problems that are nowadays still very difficult or even impossible to solve. Consequently, solving optimal control problems involving PDE constraints is a major challenge and considering multiple criteria further increases the complexity.

To overcome the problem of expensive function evaluations, model-order reduction is a widely used concept. Here, the underlying PDE is replaced by a surrogate model which can be solved much faster [22, 24]. In this context, reduced-order models (ROMs) based on Galerkin projection and *Proper Orthogonal Decomposition (POD)* [15] have proven to be a powerful tool, in particular in a multiquery context such as parameter estimation, uncertainty quantification or optimization (see, e.g., [12, 23]). During the last years, the first publications addressing PDE-constrained problems with multiple criteria have appeared. In [2], the model has been treated as a black box and evolutionary algorithms were applied. A comparison of different algorithms for multiobjective optimal control of the Navier–Stokes equations is presented in [21], and approaches using rigorous error analysis can be found in [3, 4, 16, 17], for instance.

In this work, we combine the results from [20, 23] in order to develop a global, derivative-free algorithm for PDE-constrained multiobjective optimization problems based on POD-ROMs. To this end, the subdivision algorithm for inexact function values presented in [20] is combined with a u-local reduced basis approach (see, e.g., [8, 13]) and error estimates for the objectives. The chapter is structured in the following way: In Sect. 2, the PDE-constrained multiobjective optimization problem is introduced along with the (gradient-free version of the) subdivision algorithm. In Sect. 3, an a posteriori error estimator for the individual objectives is derived. In Sect. 4, numerical results concerning both the error estimator and the overall algorithm are presented. Finally, we end with a conclusion and an outlook in Sect. 5.

Notation. If $x^1, x^2 \in \mathbb{R}^n$ are two vectors, we write $x^1 \leq x^2$ if $x_i^1 \leq x_i^2$ for $i = 1, ..., n$, and similarly for $x^1 < x^2$.

2 The Multiobjective Optimal Control Problem

Throughout this chapter, let $\Omega \subset \mathbb{R}^d$ be a bounded Lipschitz domain with boundary Γ. Further, let $(0, T) \subset \mathbb{R}$ be a time interval, $Q := (0, T) \times \Omega$ and $\Sigma := (0, T) \times \Gamma$. The spatial domain contains subdomains $\Omega_\nu \subset \Omega$, and indicator functions are defined by $\chi_\nu(x) = 1$ if $x \in \Omega_\nu$ and $\chi_\nu(x) = 0$ otherwise ($\nu = 1, ..., m$). The initial condition y_0 belongs to $L^2(\Omega)$, and the finite-dimensional control space is given by $\mathcal{U} = \mathbb{R}^m$. We consider the following *Multiobjective Optimal Control Problem* (MOCP):

$$
\min_{u \in \mathcal{U}} J(y, u) = \frac{1}{2} \begin{pmatrix} \int_\Omega |y(T, x) - y_{d,1}(x)|^2 \, dx \\ \int_\Omega |y(T, x) - y_{d,2}(x)|^2 \, dx \\ |u|_2^2 \end{pmatrix} \tag{1a}
$$

subject to (s.t.) the semilinear PDE constraints

$$
y_t(t, x) - \Delta y(t, x) + y^3(t, x) = \sum_{\nu=1}^m u_\nu \chi_\nu(x) \quad \text{for } (t, x) \in Q,
$$
$$
\frac{\partial y}{\partial n}(t, s) = 0 \qquad\qquad \text{for } (t, s) \in \Sigma, \tag{1b}
$$
$$
y(0, x) = y_0(x) \qquad\qquad \text{for } x \in \Omega
$$

and the bilateral control constraints

$$
u_a \leq u \leq u_b \quad \text{in } \mathcal{U}. \tag{1c}
$$

In (1a), the functions $y_{d,1}$ and $y_{d,2} \in L^2(\Omega)$ are two conflicting desired states. Moreover, $|\cdot|_2$ denotes the Euclidean norm. The state variable y is given as the solution to the semilinear heat equation (1b) which we will call the *state equation* from now on. It will be shown in Sect. 2 that such a solution always exists in the weak sense, is unique and in particular belongs to $C([0, T]; L^2(\Omega))$, meaning that the integrals in (1a) are well-defined. In (1c), the control variable u is bounded by bilateral constraints u_a, $u_b \in \mathcal{U}$ with $u_a \leq u_b$. Therefore, we define the *admissible set*

$$\mathcal{U}_{ad} = \big\{ u \in \mathcal{U} \,\big|\, u_a \leq u \leq u_b \text{ in } \mathcal{U} \big\}.$$

A mathematically precise formulation of (1) follows below.

The State Equation and Its Galerkin Discretization

Let us introduce the Gelfand triple $V \hookrightarrow H = H' \hookrightarrow V'$, where $V = H^1(\Omega)$, $H = L^2(\Omega)$ and each embedding is continuous and dense. In V, we utilize the standard inner product

$$\langle \varphi, \phi \rangle_V = \int_\Omega \varphi\phi + \nabla\varphi \cdot \nabla\phi \, dx \quad \text{for } \varphi, \phi \in V.$$

In particular, we have $\|\varphi\|_H \leq \|\varphi\|_V$ for all $\varphi \in V$. We define the solution space $\mathcal{Y} = W(0, T) \cap L^\infty(Q)$ with

$$W(0, T) := L^2(0, T; V) \cap H^1(0, T; V').$$

It is well-known [28] that $W(0, T)$ together with the common inner product is a Hilbert space and continuously embeds into $C([0, T]; H)$. Next, we specify what is meant by a *solution* to (1b):

Definition 1 A function $y \in \mathcal{Y}$ is a *weak solution* to (1b) if it holds for every $\varphi \in V$:

$$\langle y_t(t), \varphi \rangle_{V' \times V} + \int_\Omega \big(\nabla y(t) \cdot \nabla\varphi + y(t)^3 \varphi\big) \, dx = \sum_{\nu=1}^m u_\nu \int_{\Omega_\nu} \varphi \, dx \text{ æin } (0, T), \tag{2}$$

$$\int_\Omega y(0)\varphi \, dx = \int_\Omega y_0\varphi \, dx.$$

Here, 'a.e.' stands for 'almost everywhere'. Existence of a unique solution to (2) follows from [28, Theorems 5.5 and 5.8].

System (2) represents a nonlinear problem posed in infinite-dimensional function spaces. It is solved in practice by a discretization method. Given linearily independent spatial basis functions $\varphi_1, \ldots, \varphi_n \in V$, the space V is replaced by an n-dimensional subspace

$$V^h = \text{span}\{\varphi_1, \ldots, \varphi_n\}.$$

We endow V^h with the V-topology. Additionally, the time interval $(0, T)$ is replaced by a time grid $0 = t_1 < \ldots < t_{n_t} = T$ with trapezoidal weights $\beta_1, \ldots, \beta_{n_t} > 0$. For the sake of readability, we assume this to be equidistant, i.e., $t_j = (j - 1)\Delta t$, $j = 1, \ldots, n_t$ with the step size $\Delta t = T/(n_t - 1)$. Our discrete solution space is then given by

$$\mathcal{Y}^h = \Big\{ y^h = \{y_j^h\}_{j=1}^{n_t} \,\Big|\, y_j^h \in V^h \text{ for } j = 1, \ldots, n_t \Big\}$$

with the inner product $\langle y^h, \tilde{y}^h \rangle_{\mathcal{Y}^h} = \sum_{j=1}^{n_t} \beta_j \langle y_j^h, \tilde{y}_j^h \rangle_V$ for $y^h, \tilde{y}^h \in \mathcal{Y}^h$. A Galerkin method is employed to replace the infinite-dimensional problem (2) with a finite-dimensional version. Typically, n is a very large number which is why we refer to the solution of the resulting system as a *high-fidelity solution*:

Definition 2 A function $y^h \in \mathcal{Y}^h$ is called a *high-fidelity (HF) solution* to (1b) if it holds for every $\varphi^h \in V^h$:

$$
\int_\Omega \frac{y_j^h - y_{j-1}^h}{\Delta t} \varphi^h + \nabla y_j^h \cdot \nabla \varphi^h + (y_j^h)^3 \varphi^h \, dx = \sum_{v=1}^m u_v \int_{\Omega_v} \varphi^h \, dx, \quad j = 2, ..., n_t,
$$

$$
\int_\Omega y_1^h \varphi^h \, dx = \int_\Omega y_0 \varphi^h \, dx.
$$

(3)

Remark 1 (a) To solve (3) numerically we proceed as follows: First, we introduce the *mass matrix* $\mathbf{M} \in \mathbb{R}^{n \times n}$ and *stiffness matrix* $\mathbf{A} \in \mathbb{R}^{n \times n}$ along with the *control matrix* $\mathbf{B} \in \mathbb{R}^{n \times m}$ by

$$
\mathbf{M}_{il} = \int_\Omega \varphi_i \varphi_l \, dx, \quad \mathbf{A}_{il} = \int_\Omega \nabla \varphi_i \cdot \nabla \varphi_l \, dx, \quad \mathbf{B}_{iv} = \int_{\Omega_v} \varphi_i \, dx
$$

for $1 \leq i, l \leq n$ and $1 \leq v \leq m$. Here, $\varphi_1, ..., \varphi_\ell$ is a basis of V^h. Moreover, we define the nonlinear mapping $\mathbf{N} : \mathbb{R}^n \to \mathbb{R}^n$ as

$$
\mathbf{N}(y) = \left(\int_\Omega \left(\sum_{\mu=1}^n y_\mu \varphi_\mu \right)^3 \varphi_l \, dx \right)_{1 \leq l \leq n} \quad \text{for } y = (y_\mu)_{1 \leq \mu \leq n}
$$

and the vector

$$
y_0 = \left(\int_\Omega y_0 \varphi_i \, dx \right)_{1 \leq i \leq n}.
$$

Then, we make the following Galerkin ansatz for the HF solution:

$$
y_j^h(x) = \sum_{i=1}^n Y_{ij} \varphi_i(x), \quad x \in \Omega,
$$

with the HF coefficient matrix $\mathbf{Y} = ((Y_{ij}))_{1 \leq i \leq n, 1 \leq j \leq n_t}$. Now, the HF system (3) is expressed as the following nonlinear algebraic system:

$$
(\mathbf{M} + \Delta t \mathbf{A}) \mathbf{Y}_{\cdot, j} + \Delta t \mathbf{N}(\mathbf{Y}_{\cdot, j}) = \mathbf{M} \mathbf{Y}_{\cdot, j-1} + \Delta t \mathbf{B} u, \quad j = 2, ..., n_t,
$$

$$
\mathbf{M} \mathbf{Y}_{\cdot, 1} = y_0.
$$

(4)

where $\mathbf{Y}_{\cdot, j} \in \mathbb{R}^n$ is the jth column vector of the matrix $\mathbf{Y}$.

(b) Of course, a discretization error occurs between solutions of (2) and (3). How-
    ever, this error is not considered in the present paper, but our main focus is on a
    sufficiently accurate model-order reduction, for which it is common practice to
    accept the HF solution as the truth solution.   ◊

We suppose that there exists a continuous solution operator

$$S^h : \mathcal{U} \to \mathcal{Y}^h, \quad u \mapsto y^h = S^h u, \text{ where } (y^h, u) \text{ solves (3)}.$$

The definition of the control-to-state operator S^h and the admissible set $\mathcal{U}_{ad}$ allows
us to rewrite (1) as a minimization problem in the control variable only:
We introduce the *cost function*

$$\hat{J}^h : \mathcal{U}_{ad} \to \mathbb{R}^3, \quad \hat{J}^h(u) = J(S^h u, u) \text{ for } u \in \mathcal{U}_{ad}$$

and the mathematical realization of (1) which we will call the *reduced MOCP*:

$$\min \quad \hat{J}^h(u) \quad \text{s.t.} \quad u \in \mathcal{U}_{ad}. \tag{$\hat{\mathbf{P}}^h$}$$

In the sequel, we suppose that (1) possesses a (locally) optimal solution $\bar{u}^h \in \mathcal{U}_{ad}$.

Multiobjective Optimization

Consider the general multiobjective optimization problem

$$\min_{u \in \mathcal{U}_{ad}} \hat{J}^h(u) = \min_{u \in \mathcal{U}_{ad}} \begin{pmatrix} \hat{J}_1^h(u) \\ \vdots \\ \hat{J}_k^h(u) \end{pmatrix}, \tag{5}$$

where $\hat{J}^h : \mathcal{U}_{ad} \to \mathbb{R}^k$ is a vector-valued objective function with continuously dif-
ferentiable scalar objective functions $\hat{J}_i^h : \mathcal{U}_{ad} \to \mathbb{R}$ for $i = 1, \ldots, k$. The space of
the 'parameters' u is called the *decision space*, and the function $\hat{J}^h$ is a mapping
to the k-dimensional *objective space*. In contrast to single objective optimization
problems, there exists no total order of the objective function values in $\mathbb{R}^k$ for $k \geq 2$.
Consider, for example, the points $u_1 = (3, 2)$ and $u_2 = (1, 3)$ and $u_3 = (2, 1)$. Then
neither $u_2 < u_1$ nor $u_1 < u_2$. However, $u_3 < u_1$ since $u_{3,1} < u_{1,1}$ and $u_{3,2} < u_{1,2}$.
    A consequence of the lack of a total order is that we cannot expect to find isolated
optimal points. Instead, the solution of (5) is the set of optimal compromises, the
so-called *Pareto set*:

Definition 3

(a) A point $u^* \in \mathcal{U}_{ad}$ *dominates* a point $u \in \mathcal{U}_{ad}$, if $\hat{J}^h(u^*) \leq \hat{J}^h(u)$ and $\hat{J}^h(u^*) \neq \hat{J}^h(u)$.

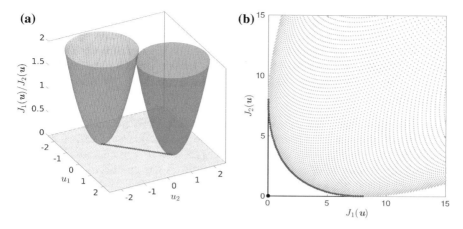

Fig. 1 Red lines: Sketch of a Pareto set (**a**) and Pareto front (**b**) for a multiobjective optimization problem of the form (5) with $m = 2$ and $k = 2$. The objectives are to minimize the two functions represented by the blue and the orange paraboloid, respectively, where the third axis represents the function values

(b) A point $u^* \in \mathcal{U}_{\text{ad}}$ is called *(globally) Pareto optimal* if there exists no point $u \in \mathcal{U}_{\text{ad}}$ dominating u^*. The image $\hat{J}^h(u^*)$ of a (globally) Pareto optimal point u^* is called a *(globally) Pareto optimal value*.

(c) The set of non-dominated points is called the *Pareto set* $\mathcal{P}_S \subset \mathbb{R}^m$, its image with respect to $\hat{J}^h$ the *Pareto front* $\mathcal{P}_F \subset \mathbb{R}^k$.

(d) When comparing sets, a set $\mathcal{B}^* \subset \mathcal{U}_{\text{ad}}$ *dominates* a set $\mathcal{B} \subset \mathcal{U}_{\text{ad}}$ if for every point $u \in \mathcal{B}$, there exists at least one point $u^* \in \mathcal{B}^*$ dominating u.

Consequently, for each point that is contained in the Pareto set, one can only improve one objective by accepting a trade-off in at least one other objective (cf. Fig. 1b). A more detailed introduction to multiobjective optimization can be found in [10, 18], for instance.

As mentioned in the introduction, many different algorithms for solving MOPs exist. In this work, we focus on the set-oriented methods introduced in [7]. In the following, we will shortly describe the gradient-free approach. The numerical realization of this technique motivates the use of u-local reduced basis approaches.

Gradient-Free Multiobjective Optimization

A straightforward method to compute the global Pareto set of an MOCP is to directly make use of the non-dominance property d). Based on this property, in a set of points one can determine the subset of non-dominated points by comparing the objective values.

Since the solution of $(\hat{\mathbf{P}}^h)$ is a set, this motivates the use of set-oriented approaches. A method of this type which computes an outer covering of the Pareto set was proposed in [7]. In the resulting *subdivision algorithm*, we begin with one initial covering $\mathscr{B}_0 = \mathfrak{U}_{\text{ad}}$ of the control set. This covering is alternatingly refined by *subdivision* and then reduced to the subset containing the Pareto set by *selection* until a sufficient level of refinement is achieved. This way, the algorithm yields a nested sequence of increasingly refined coverings $\mathscr{B}_0, \mathscr{B}_1, \ldots$ of the entire Pareto set $\mathscr{P}_S$, where each $\mathscr{B}_s$ is a subset of $\mathscr{B}_{s-1}$.

Remark 2 In the numerical realization, the elements of $\mathscr{B}_s$ are generalized rectangles B (from now on referred to as *boxes*). These are represented by a finite number of sample points, see Fig. 2 for an illustration. Those sample points are distributed randomly or on an equidistant grid, for example. In the numerical evaluation of the set-valued non-dominance property (cf. Definition 3d)), a box is considered as dominated if all sample points are dominated. The treatment of constraints is handled similarly: A box B violates the constraints if $u \notin \mathfrak{U}_{\text{ad}}$ for all sample points $u \in B$. $\lozenge$

In the subdivision step, we construct a new collection of subsets $\hat{\mathscr{B}}_s$ from $\mathscr{B}_{s-1}$ such that

$$\bigcup_{B \in \hat{\mathscr{B}}_s} B = \bigcup_{B \in \mathscr{B}_{s-1}} B, \quad \text{diam}(\hat{\mathscr{B}}_s) = \theta_s \text{diam}(\mathscr{B}_{s-1}), \quad 0 < \theta_{min} \leq \theta_s \leq \theta_{max} < 1,$$

for some $\theta_{min}, \theta_{max} \in [0, 1]$. Here, $\text{diam}(\mathscr{B}_s)$ is the *box diameter*:

$$\text{diam}(\mathscr{B}_s) = \max_{B \in \mathscr{B}_s} \text{diam}(B).$$

This means that the box size is reduced in each subdivision step. In practice, the subdivision is performed by consecutive bisection of the boxes contained in the current covering, cyclically with respect to the coordinate directions.

In the selection step, all dominated boxes B are discarded (cf. Remark 2 for the numerical realization). The dominated points are identified by evaluating the objective function for the sample points of all boxes B and then using a non-dominance test [25]. If additional constraints are present, then prior to applying the non-dominance test, all boxes violating the constraints are discarded, i.e., boxes B where $u \notin \mathfrak{U}_{\text{ad}}$ for all sample points $u \in B$. However, we here only consider box constraints which are automatically satisfied by appropriately choosing the initial covering $\mathscr{B}_0$. The gradient-free subdivision algorithm is summarized in Algorithm 1 and visualized in Fig. 2.

Besides globality, a benefit of this technique is that it can easily be applied to higher dimensions, whereas in particular geometric approaches struggle with a larger number of objectives. However, the computational cost increases exponentially with the dimension of the Pareto set such that in practice, we are restricted to a moderate number of objectives, i.e., $k \leq 5$.

We can conclude that in order to solve $(\hat{\mathbf{P}}^h)$ by Algorithm 1, we have to evaluate the objectives $\hat{J}^h$ for many different controls u which can quickly result in a prohibitively large computational effort if model evaluations are expensive. The idea is therefore

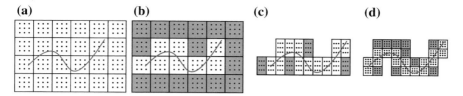

Fig. 2 Global subdivision algorithm. The Pareto set is shown in red, the sample points are indicated by black dots. **a** Sampling of boxes B in $\mathcal{B}_s$. **b** Selection step for $\mathcal{B}_s$. The gray boxes are dominated and hence discarded. **c** Selection step for $\mathcal{B}_{s+1}$. **d** Selection step for $\mathcal{B}_{s+2}$

to introduce a reduced-order model which is less expensive to solve but on the other hand only yields approximations of the exact function values.

Inexact Function Values

Suppose now that we only have approximations $\tilde{J}_i^h(u)$ of the objectives $\hat{J}_i^h(u)$ for $i = 1, \ldots, k$. To be more precise, we assume that

$$|\tilde{J}_i^h(u) - \hat{J}_i^h(u)| \leq \Delta_i^J(u) \quad \text{for } i = 1, \ldots, k, \tag{6}$$

Algorithm 1 (Subdivision algorithm)

Let $\mathcal{B}_0$ be an initial collection of finitely many subsets of the compact set $\mathcal{U}_0 = \mathcal{U}_{ad}$ such that $\bigcup_{B \in \mathcal{B}_0} B = \mathcal{U}_0$. Then, $\mathcal{B}_s$ is inductively obtained from $\mathcal{B}_{s-1}$ in two steps:

(i) Subdivision. Construct from $\mathcal{B}_{s-1}$ a new collection of subsets $\hat{\mathcal{B}}_s$ such that

$$\bigcup_{B \in \hat{\mathcal{B}}_s} B = \bigcup_{B \in \mathcal{B}_{s-1}} B, \quad \text{diam}(\hat{\mathcal{B}}_s) = \theta_s \text{diam}(\mathcal{B}_{s-1}), \quad 0 < \theta_{min} \leq \theta_s \leq \theta_{max} < 1.$$

(ii) Selection. Define the new collection $\mathcal{B}_s$ by

$$\tilde{\mathcal{B}}_s = \left\{ B \in \hat{\mathcal{B}}_s \;\middle|\; \exists u \in \hat{\mathcal{B}}_s \text{ with } u \in \mathcal{U}_{ad} \right\}, \tag{Constraints}$$

$$\mathcal{B}_s = \Big\{ B \in \tilde{\mathcal{B}}_s \;\Big|\; \nexists \hat{B} \in \tilde{\mathcal{B}}_s \text{ such that } \hat{B}$$

$$\text{dominates } B \text{ according to Definition 3 d)} \Big\}. \tag{Non-Dominance}$$

---

where the upper bounds $\Delta_i^J(u)$ can be evaluated more efficiently than an actual evaluation of $\hat{J}^h$. For a more detailed introduction to multiobjective optimization with uncertainties, also with respect to the treatment of inexact gradients, the reader is referred to [20]. In the section "Error Estimation", we will introduce the particular bound that is used for the reduced cost function $\hat{J}^h$ in $(\hat{\mathbf{P}}^h)$.

Based on the inexactness, we now extend the concept of *non-dominance* (cf. Definition 3) to inexact function values:

Definition 4 Consider the multiobjective optimization problem (5), where the objective functions $\hat{J}_i^h(u)$, $i = 1, \ldots, k$, are only known approximately according to (6).

(a) A point $u^* \in \mathcal{U}_{\mathsf{ad}}$ *confidently dominates* a point $u \in \mathcal{U}_{\mathsf{ad}}$, if $\tilde{J}_i^h(u^*) + \Delta_i^J(u) \leq \tilde{J}_i^h(u) - \Delta_i^J(u)$ for $i = 1, \ldots, k$ and $\tilde{J}_i^h(u^*) + \Delta_i^J(u) < \tilde{J}_i^h(u) - \Delta_i^J(u)$ for at least one $i \in 1, \ldots, k$.
(b) A set $\mathcal{B}^* \subset \mathcal{U}_{\mathsf{ad}}$ *confidently dominates* a set $\mathcal{B} \subset \mathcal{U}_{\mathsf{ad}}$ if for every point $u \in \mathcal{B}$, there exists at least one point $u^* \in \mathcal{B}^*$ confidently dominating u.
(c) The *set of almost non-dominated points*, which is a superset of the Pareto set $\mathcal{P}_S$, is defined as:

$$\mathcal{P}_{S,\Delta} = \left\{ u^* \in \mathcal{U}_{\mathsf{ad}} \middle| \nexists u \in \mathcal{U}_{\mathsf{ad}} \text{ with } \tilde{J}_i^h(u) + \Delta_i^J(u) \leq \tilde{J}_i^h(u^*) - \Delta_i^J(u), \right.$$
$$\left. \text{for } i = 1, \ldots, k \right\}. \tag{7}$$

Using Definition 4, we can extend Algorithm 1 to the situation where the function values are only known approximately. This can simply be achieved by changing the non-dominance test in the selection step to the stronger condition in (7). This allows us to compute the set of almost non-dominated points $\mathcal{P}_{S,\Delta}$ based on reduced-order models using Algorithm 2. Consequently, if we are interested in approximating the exact Pareto set with a prescribed accuracy, we are faced with the challenge to control the error of the underlying reduced-order model such that it satisfies condition (6). Algorithm 2 requires a large number of function evaluations for different controls u with an error as small as possible. This task can be addressed by using u-local reduced basis approaches [8, 13]. Instead of building one model which is globally valid, the

---

Algorithm 2 (Inexact subdivision algorithm)

Let $\mathcal{B}_0$ be an initial collection of finitely many subsets of the compact set $\mathcal{U}_0 = \mathcal{U}_{\mathsf{ad}}$ such that $\bigcup_{B \in \mathcal{B}_0} B = \mathcal{U}_0$. Then, $\mathcal{B}_s$ is inductively obtained from $\mathcal{B}_{s-1}$ in two steps:

(i) Subdivision. Construct from $\mathcal{B}_{s-1}$ a new collection of subsets $\hat{\mathcal{B}}_s$ such that

$$\bigcup_{B \in \hat{\mathcal{B}}_s} B = \bigcup_{B \in \mathcal{B}_{s-1}} B, \quad \mathrm{diam}(\hat{\mathcal{B}}_s) = \theta_s \mathrm{diam}(\mathcal{B}_{s-1}), \quad 0 < \theta_{min} \leq \theta_s \leq \theta_{max} < 1.$$

(ii) Selection. Define the new collection $\mathcal{B}_s$ by

$$\widetilde{\mathcal{B}}_s = \left\{ B \in \hat{\mathcal{B}}_s \middle| \exists u \in \hat{\mathcal{B}}_s \text{ with } u \in \mathcal{U}_{\mathsf{ad}} \right\}, \tag{Constraints}$$

$$\mathcal{B}_s = \left\{ B \in \widetilde{\mathcal{B}}_s \middle| \nexists \hat{B} \in \widetilde{\mathcal{B}}_s \text{ such that } \hat{B} \text{ confidently} \right.$$
$$\left. \text{dominates } B \text{ according to Definition 4 d)} \right\}. \tag{Non-Dominance}$$

---

concept there is to store several, locally valid ROMs in a library and evaluate the objective functions with the model with the best accuracy. This has the advantage that the respective models can be smaller in size, whereas a reduced model which is accurate within a large range of controls may become too high-dimensional to possess the necessary efficiency. This advantage comes with the price that many high-fidelity function evaluations need to be performed in order to build the library of different ROMs. Note that the local validity mentioned here relates to the control domain. Localized reduced basis approaches for the state space have been treated in [1, 19].

3 Model-Order Reduction

Algorithm 1 implicitly presents us with the task of evaluating the cost function $\hat{J}^h(u)$ for all sample points $u \in B$ and all sets $B \in \mathscr{B}_s$. Each of these evaluations requires the system (2) (respectively (3) in the numerical implementation) to be solved for the current control. As stated before, it is reasonable in this multiquery context to apply model-order reduction techniques in order to reduce the computational effort for the optimization.

The POD Galerkin Scheme for the State Equation

In this work, we utilize the POD method to compute the ROMs; cf. [15]. Suppose that we have chosen an admissible control $u \in \mathcal{U}_{\text{ad}}$ and we would like to build a u-local surrogate model which is highly accurate for the data associated with this control. Let $y^h = \mathcal{S}^h u$ denote the associated solution to (3). Then we consider the linear space of snapshots

$$\mathcal{V}^h = \text{span}\left\{ y_j^h \mid j = 1, ..., n_t \right\} \subset V^h \subset V \quad \text{with } \mathfrak{d} = \dim \mathcal{V}^h \leq \min(n, n_t).$$

For any finite $\ell \leq \mathfrak{d}$, we are interested in determining a POD basis of rank ℓ which minimizes the mean square error between y^h and their corresponding ℓ-th partial Fourier sums in the resulting subspace on average over the discrete time interval

$$\begin{cases} \min \sum_{j=1}^{n_t} \beta_j \left\| y_j^h - \sum_{i=1}^{\ell} \langle y_j^h, \psi_i^h \rangle_V \, \psi_i^h \right\|_V^2 \\ \text{s.t. } \{\psi_i^h\}_{i=1}^{\ell} \subset V^h \text{ and } \langle \psi_i^h, \psi_l^h \rangle_V = \delta_{il} \text{ for } 1 \leq i, l \leq \ell. \end{cases} \tag{$\mathbf{P}^\ell$}$$

A solution $\{\psi_i^h\}_{i=1}^{\ell}$ to $(\mathbf{P}^\ell)$ is called *POD basis of rank ℓ*. Let us introduce the linear, compact, selfadjoint, and nonnegative operator $\mathscr{R}^h : V^h \to V^h$ by

$$\mathscr{R}^h \psi^h = \sum_{j=1}^{n_t} \beta_j \, \langle y_j^h, \psi^h \rangle_V \, y_j^h \quad \text{for } \psi^h \in V^h.$$

Then, it is well-known [12, Theorem 1.15] that a solution $\{\psi_i^h\}_{i=1}^\ell$ to $(\mathbf{P}^\ell)$ is given by the eigenvectors associated with the ℓ largest eigenvalues of $\mathscr{R}^h$:

$$\mathscr{R}^h \psi_i^h = \lambda_i^h \psi_i^h \text{ for } 1 \le i \le \ell, \quad \lambda_1^h \ge \ldots \ge \lambda_\ell^h \ge \lambda_{\ell+1}^h \ge \ldots \lambda_{\mathfrak{d}}^h > \lambda_{\mathfrak{d}}^h = \ldots = 0.$$

Moreover, the POD basis $\{\psi_i^h\}_{i=1}^\ell$ of rank ℓ satisfies $\psi_i^h \in V^h$ for $1 \le i \le \ell$ and

$$\sum_{j=1}^{n_t} \beta_j \left\| y_j^h - \sum_{i=1}^\ell \langle y_j^h, \psi_i^h \rangle_V \, \psi_i^h \right\|_V^2 = \sum_{i=\ell+1}^{\mathfrak{d}} \lambda_i^h.$$

Now suppose that we have computed a POD basis $\{\psi_i^h\}_{i=1}^\ell \subset V^h$ of rank $\ell \ll n$. We define the finite-dimensional subspace

$$V^{h\ell} = \text{span} \, \{\psi_1^h, \ldots, \psi_\ell^h\} \subset V^h$$

and the POD solution space

$$\mathcal{Y}^{h\ell} = \left\{ y^{h\ell} = \{y_j^{h\ell}\}_{j=1}^{n_t} \, | \, y_j^{h\ell} \in V^{h\ell} \text{ for } j = 1, \ldots, n_t \right\} \subset \mathcal{Y}^h$$

which we also endow with the inner product on $\mathcal{Y}^h$. Then, the POD solution operator $\mathcal{S}^{h\ell} : \mathcal{U} \to \mathcal{Y}^\ell$ is defined as follows: $y^{h\ell} = \mathcal{S}^{h\ell} u$ solves the following POD Galerkin scheme for every $\psi^h \in V^{h\ell}$:

$$\int_\Omega \frac{y_j^{h\ell} - y_{j-1}^{h\ell}}{\Delta t} \psi^h + \nabla y_j^{h\ell} \cdot \nabla \psi^h + (y_j^{h\ell})^3 \psi^h \, dx = \sum_{i=1}^m u_i \int_{\Omega_i} \psi^h \, dx, \quad j = 2, \ldots, n_t, \tag{8}$$

$$y_1^{h\ell} = \mathscr{P}^{h\ell} y_0,$$

where the linear projection operator $\mathscr{P}^{h\ell} : H \to V^{h\ell}$ is given by

$$y_1^{h\ell} = \mathscr{P}^{h\ell} y_0 = \arg\min_{\varphi^\ell \in V^{h\ell}} \|\varphi^\ell - y_0\|_H.$$

The choice of $\mathscr{P}^{h\ell}$ is motivated by rate of convergence results shown in [27]. Similarily to the reduced cost function $\hat{J}^h$, we introduce the *reduced-order cost function*

$$\hat{\jmath}^{h\ell} : \mathcal{U}_{\text{ad}} \to \mathbb{R}^3, \quad \hat{\jmath}^{h\ell}(u) = J(\mathcal{S}^{h\ell} u, u) \text{ for } u \in \mathcal{U}_{\text{ad}}$$

Remark 3 We can exploit the numerical structure of Problem (8) to precompute many recurring values. For the POD Galerkin system (8), let the basis vectors be given by

$$\psi_l^h(x) = \sum_{i=1}^n \Psi_{il} \varphi_i(x) \quad \text{for } x \in \Omega \text{ and } l = 1, ..., \ell.$$

We call $\Psi \in \mathbb{R}^{n \times \ell}$ the *basis matrix* and write

$$y_j^{h\ell}(x) = \sum_{l=1}^{\ell} Y_{lj}^{\ell} \psi_l^h(x) = \sum_{l=1}^{\ell} Y_{lj}^{\ell} \sum_{i=1}^n \Psi_{il} \varphi_i(x) = \sum_{i=1}^n \Psi_{i,\cdot} Y_{\cdot,j}^{\ell} \varphi_i(x)$$

for $x \in \Omega$ and $j = 1, ..., n_t$, where $Y_{\cdot,j}^{\ell} \in \mathbb{R}^{\ell}$ denotes the jth column of the coefficient matrix $Y^{\ell} \in \mathbb{R}^{n \times \ell}$. With this, (8) can be written as the nonlinear algebraic system:

$$\left(M^{\ell} + \Delta t A^{\ell} \right) Y_{\cdot,j}^{\ell} + \Delta t N^{\ell}(Y_{\cdot,j}^{\ell}) = M^{\ell} Y_{\cdot,j-1}^{\ell} + \Delta t B^{\ell} u \quad \text{for } j = 2, ..., n_t,$$

$$M^{\ell} Y_{\cdot,1}^{\ell} = y_0^{\ell}$$

with

$$M^{\ell} = \Psi^{\top} M \Psi \in \mathbb{R}^{\ell \times \ell}, \qquad\qquad A^{\ell} = \Psi^{\top} A \Psi \in \mathbb{R}^{\ell \times \ell},$$

$$B^{\ell} = \Psi^{\top} B \in \mathbb{R}^{\ell \times m}, \qquad\qquad y_0^{\ell} = \Psi^{\top} M y_0 \in \mathbb{R}^{\ell}.$$

and the nonlinear mapping $N^{\ell} : \mathbb{R}^{\ell} \to \mathbb{R}^{\ell}$ defined as

$$N^{\ell}(y^{\ell}) = \left(\int_{\Omega} \left(\sum_{\mu=1}^{\ell} y_{\mu}^{\ell} \psi_{\mu}^h \right)^3 \psi_l^h \, dx \right)_{1 \leq l \leq \ell} \quad \text{for } y^{\ell} = \left(y_{\mu}^{\ell} \right)_{1 \leq \mu \leq \ell}.$$

By precomputing the occurring matrices above once in an *offline phase* and calling upon them in the *online solving phase*, the computational time can be drastically improved. However, the nonlinear terms $N^{\ell}(Y_{\cdot,j}^{\ell})$, $j = 2, ..., n_t$, still need to be computed in the HF space of dimension n. In order to remedy this, we employ a standard Discrete Empirical Interpolation Method (DEIM) with proper offline-/online decomposition to circumvent this bottleneck. For details on DEIM, we refer the interested reader to [5], for example.   ◇

Error Estimation

In this section, we will present an a posteriori error estimator for the state equation for an arbitrarily chosen admissible control $u \in \mathcal{U}_{ad}$ which was not necessarily used to build the reduced-order model; cf. [11, 23].

Theorem 1 *Let a finite-dimensional subspace $V^{h\ell}$ be given as described in the previous section and $u \in \mathcal{U}_{ad}$ an arbitrary admissible control. Define the state and reduced state solutions as $y^h = \mathcal{S}^h u$ and $y^{h\ell} = \mathcal{S}^{h\ell} u$. Then the following a posteriori error estimate*

$$\|y_j^h - y_j^{h\ell}\|_H^2 \leq \Delta_j^{\mathrm{pr}}(u) \quad \text{for } j = 2, \ldots, n_t \tag{9}$$

holds with the a posteriori estimator

$$\Delta_j^{\mathrm{pr}}(u) = \left(1 - \Delta t\right)^{1-j} \|y_1^h - y_1^{h\ell}\|_H^2 + \Delta t \sum_{\mu=2}^{j} \left(\left(1 - \Delta t\right)^{\mu-j-1} \|R_\mu^{h\ell}(u)\|_{(V^h)'}^2\right) \tag{10}$$

for $\Delta t \in (0, 1)$ and

$$\|R_j^{h\ell}(u)\|_{(V^h)'} = \max_{\varphi^h \in V^h \setminus \{0\}} \frac{\left|\langle R_j^{h\ell}(u), \varphi^h \rangle_{(V^h)', V^h}\right|}{\|\varphi^h\|_V}.$$

The residual term $R_j^{h\ell}(u)$ is defined for $j = 1, ..., n_t$ and $\varphi^h \in V^h$ as:

$$\langle R_j^{h\ell}(u), \varphi^h \rangle_{(V^h)', V^h} = \int_\Omega \frac{y_j^{h\ell} - y_{j-1}^{h\ell}}{\Delta t} \varphi^h + \nabla y_j^{h\ell} \cdot \nabla \varphi^h + (y_j^{h\ell})^3 \varphi^h \, dx$$

$$- \sum_{\nu=1}^{m} u_\nu \int_\Omega \chi_\nu \varphi^h \, dx \quad \text{for } \varphi^h \in V^h.$$

Proof The proof follows similar arguments as for the linear heat equation. By defining the error $e_j^{h\ell} = y_j^h - y_j^{h\ell}$, we obtain for every $\varphi^h \in V^h$ and $j = 2, ..., n_t$:

$$\left\langle \frac{e_j^{h\ell} - e_{j-1}^{h\ell}}{\Delta t}, e_j^{h\ell} \right\rangle_H + \int_\Omega |\nabla e_j^{h\ell}|_2^2 + \left[(y_j^h)^3 - (y_j^{h\ell})^3\right] e_j^{h\ell} \, dx = \langle R_j^{h\ell}(u), e_j^{h\ell} \rangle_{(V^h)', V^h}.$$

By using the mean value theorem, we find that

$$\int_\Omega \left[(y_j^h)^3 - (y_j^{h\ell})^3\right] e_j^{h\ell} \, dx = 3 \int_\Omega \left(s y_j^h + (1-s) y_j^{h\ell}\right)^2 |e_j^{h\ell}|^2 \, dx \geq 0.$$

Here, $s \in (0, 1)$ is the parameter indicating the mean value on the line between y_j^h and $y_j^{h\ell}$. Moreover, we have

$$\langle e_j^{h\ell} - e_{j-1}^{h\ell}, e_j^{h\ell} \rangle_H = \frac{1}{2} \|e_j^{h\ell} - e_{j-1}^{h\ell}\|_H^2 + \frac{1}{2} \|e_j^{h\ell}\|_H^2 - \frac{1}{2} \|e_{j-1}^{h\ell}\|_H^2,$$

$$\int_\Omega |\nabla e_j^{h\ell}|_2^2 \, dx = \|e_j^{h\ell}\|_V^2 - \|e_j^{h\ell}\|_H^2.$$

Therefore, we conclude from Young's inequality that

$$\frac{1}{2}\|e_j^{h\ell}\|_H^2 + \Delta t\left(\|e_j^{h\ell}\|_V^2 - \|e_j^{h\ell}\|_H^2\right) \leq \frac{1}{2}\|e_{j-1}^{h\ell}\|_H^2 + \frac{\Delta t}{2}\|R_j^{h\ell}(u)\|_{(V^h)'}^2 + \frac{\Delta t}{2}\|e_j^{h\ell}\|_V^2$$

which implies

$$\|e_j^{h\ell}\|_H^2 + \Delta t\left(\|e_j^{h\ell}\|_V^2 - 2\|e_j^{h\ell}\|_H^2\right) \leq \|e_{j-1}^{h\ell}\|_H^2 + \Delta t\|R_j^{h\ell}(u)\|_{(V^h)'}^2.$$

Hence,

$$\left(1 - \Delta t\right)\|e_j^{h\ell}\|_H^2 \leq \|e_j^{h\ell}\|_H^2 + \Delta t\left(\|e_j^{h\ell}\|_V^2 - 2\|e_j^{h\ell}\|_H^2\right) \leq \|e_{j-1}^{h\ell}\|_H^2 + \Delta t\|R_j^{h\ell}(u)\|_{(V^h)'}^2.$$

Consequently, we have for $\Delta t < 1$

$$\|e_j^{h\ell}\|_H^2 \leq \left(1 - \Delta t\right)^{-1}\|e_{j-1}^{h\ell}\|_H^2 + \Delta t\left(1 - \Delta t\right)^{-1}\|R_j^{h\ell}(u)\|_{(V^h)'}^2$$

$$\leq \left(1 - \Delta t\right)^{-2}\|e_{j-2}^{h\ell}\|_H^2 + \Delta t\sum_{\mu=j-1}^{j}\left(1 - \Delta t\right)^{\mu-j-1}\|R_\mu^{h\ell}(u)\|_{(V^h)'}^2.$$

By summation on j the claim follows. $\qquad\square$

Lemma 1 *Given the ansatz $y_j^{h\ell} = \sum_{i=1}^{\ell} Y_{ij}^{\ell}\psi_i^h$, the dual norm of the residual $R_j^{h\ell}(u)$ can be computed as follows for $j = 2, \dots, n_t$:*

$$\|R_j^{h\ell}(u)\|_{(V^h)'} = \sqrt{d_j^T(M + A)^{-1}d_j}$$

where

$$d_j = M\Psi\frac{Y_{\cdot,j}^{\ell} - Y_{\cdot,j-1}^{\ell}}{\Delta t} + A\Psi Y_{\cdot,j}^{\ell} + N(\Psi Y_{\cdot,j}^{h\ell}) - Bu \in \mathbb{R}^{N_x}.$$

Proof For an arbitrarily given $\varphi^h \in V^h$, we write $\varphi^h = \sum_{i=1}^{n} b_i\varphi_i$ with $b \in \mathbb{R}^n$ to obtain for $j = 2, \dots, n_t$:

$$\langle R_j^{h\ell}(u), \varphi^h\rangle_{(V^h)',V^h} = b^T d_j = b^T(M + A)(M + A)^{-1}d_j = \langle \widetilde{R}_j^{h\ell}(u), \varphi\rangle_V$$

where $\widetilde{R}_j^{h\ell}(u) = \sum_{i=1}^{n}\widetilde{d}_j\varphi_i \in V^h$ is the Riesz representative of $R_j^{h\ell}$ with $\widetilde{d}_j = (M + A)^{-1}d_j$. Therefore, it is

$$\|R_j^{h\ell}(u)\|_{(V^h)'} = \|\widetilde{R}_{h\ell}^{j}(u)\|_V = \sqrt{\widetilde{d}_j^T(M + A)\widetilde{d}_j} = \sqrt{d_j^T(M + A)^{-1}d_j}$$

which gives the claim. $\qquad\square$

The above representation of $d_j \in \mathbb{R}^{N_x}$ is a vector belonging to the high-fidelity model which to be computed in order to obtain the error estimator Δ^{pr}, whereas the matrices $M\Psi$, $A\Psi \in \mathbb{R}^{N_x \times \mathbb{R}^\ell}$ can be precomputed in an offline phase, the evaluation of the nonlinearity $N(\Psi Y_{\cdot,j}^{h\ell})$ is of full order and has to be performed during the online phase. It is a price to be paid for the application of error estimation: Instead of solving the nonlinear system (4) using a Newton method for each control occurring during the optimization, we solve the low-dimensional system (8) with a Newton method and additionally evaluate the error estimator (10). This usually still results in a reduced overall computation time. As an indication, for the numerical setup from Sect. 4, a high-fidelity solve takes on average 0.31 s, a reduced-order solve 0.05 s, and the evaluation of the error estimator 0.07 s, so the computational effort is reduced roughly by a factor of $1/3$. One could avoid the full-order evaluations by utilizing empirical interpolation (cf. [9]), but then the presented error estimator is not rigorous.

Corollary 1 *Let $V^{h\ell}$ be introduced as in the previous section and $u \in \mathcal{U}_{ad}$ an arbitrary admissible control. Then the following a posteriori estimate for the cost function holds for $i = 1, 2$:*

$$\left| \hat{J}_i^h(u) - \hat{J}_i^{h\ell}(u) \right| \le \sqrt{2\Delta_{n_t}^{pr}(u)\hat{J}^{h\ell}(u)} + \frac{1}{2}\Delta_{n_t}^{pr}(u) =: \Delta_i^J(u), \qquad (11)$$

where $\Delta_{n_t}^{pr}(u)$ has been introduced in (10).

Proof We fix $i \in \{1, 2\}$ and observe using basic estimations:

$$\begin{aligned}
\left| \hat{J}_i^h(u) - \hat{J}_i^{h\ell}(u) \right| &= \frac{1}{2} \left| \|y_{n_t}^h - y_{d,i}\|_H^2 - \|y_{n_t}^{h\ell} - y_{d,i}\|_H^2 \right| \\
&= \frac{1}{2} \left| \|y_{n_t}^h - y_{d,i}\|_H + \|y_{n_t}^{h\ell} - y_{d,i}\|_H \right| \\
&\qquad \cdot \left| \|y_{n_t}^h - y_{d,i}\|_H - \|y_{n_t}^{h\ell} - y_{d,i}\|_H \right| \\
&\le \frac{1}{2} \left(\|y_{n_t}^h - y_{d,i}\|_H + \|y_{n_t}^{h\ell} - y_{d,i}\|_H \right) \cdot \|y_{n_t}^h - y_{n_t}^{h\ell}\|_H \\
&\le \frac{1}{2} \left(2\|y_{n_t}^h - y_{d,i}\|_H + \|y_{n_t}^h - y_{n_t}^{h\ell}\|_H \right) \cdot \|y_{n_t}^h - y_{n_t}^{h\ell}\|_H \\
&= \sqrt{2\hat{J}^{h\ell}(u)} \cdot \|y_{n_t}^h - y_{n_t}^{h\ell}\|_H + \frac{1}{2}\|y_{n_t}^h - y_{n_t}^{h\ell}\|_H^2.
\end{aligned}$$

The term $\|y_{n_t}^h - y_{n_t}^{h\ell}\|_H^2$ can be further bounded by the state estimator (9) for $j = n_t$, and this yields (11). $\qquad\square$

Remark 4 In the context of the section "Inexact Function Values", we can interpret $\tilde{J}^h(u) = \hat{J}^{h\ell}(u)$ as an approximation of $\hat{J}^h(u)$ for a given control $u \in \mathcal{U}_{ad}$. $\qquad\Diamond$

Table 1 Sample evaluation of true error, scaled and unscaled error estimator, and their efficiency. 100 randomly chosen sample controls were evaluated, 10 of which were used to perform the scaling

	$\left|\hat{j}_1^h - \hat{j}_1^{h\ell}\right|/\left|\hat{j}_1^h\right|$	$\left|\hat{j}_1^h - \hat{j}_1^{h\ell}\right|/\Delta_1^J$, unscaled	$\left|\hat{j}_1^h - \hat{j}_1^{h\ell}\right|/\Delta_1^J$, scaled
Mean over u	0.0572	0.3421	0.6418
Min over u	0.0004	0.0099	0.0154
Max over u	0.4989	0.9285	1.7608

Heuristic Error Correction

The estimator $\Delta_{n_t}^{\mathsf{pr}}(u)$ from (9) is a so-called rigorous estimator, i.e., it is mathematically guaranteed that it serves as an upper bound for the true error. As a result, it was assumed during the proof that the unknown true error always behaves according to the worst-case scenario. In a concrete application, it is therefore often observed that this estimator overshoots the true error by a roughly constant factor. Depending on the application, it may be desirable to increase the tightness of the estimator by heuristically incorporating data from a precomputed offline phase. In the context of multiobjective optimization, all points resulting in optimal compromises are contained in the Pareto set; cf. Definition 3. If the Pareto front is very 'steep' in certain regions, i.e., very small changes in one objective lead to strong changes in another objective, then small inaccuracies can result in very large deviations from the exact solution. Since this is exactly the situation in the problem considered in Sect. 4 (cf. Fig. 4c), we will introduce such a heuristic in this section. However, it has to be noted that the mathematical rigor that is prevalent at this point will be lost by doing this. We assume an over-estimation of the true error of the form

$$\Delta_{n_t}^{\mathsf{pr}}(u) \approx C_{\mathsf{sc}} \cdot \|y_{n_t}^h - y_{n_t}^{h\ell}\|_H^2, \tag{12}$$

where $C_{\mathsf{sc}} > 1$ is an unknown scaling factor. Let a finite sample set $\mathcal{U}_{\mathsf{sc}} \subset \mathcal{U}_{\mathsf{ad}}$ be given prior to the optimization. For a given reduced-order model, we compute the full and reduced state solutions $y_{n_t}^h = \mathcal{S}^h u$, $y^{h\ell} = \mathcal{S}^{h\ell} u$ for all sample controls $u \in \mathcal{U}_{\mathsf{sc}}$ and set C_{sc} to be the geometric average

$$C_{\mathsf{sc}} = \left(\prod_{u \in \mathcal{U}_{\mathsf{sc}}} \frac{\Delta_{n_t}^{\mathsf{pr}}(u)}{\|y_{n_t}^h - y_{n_t}^{h\ell}\|_H^2} \right)^{\frac{1}{|\mathcal{U}_{\mathsf{sc}}|}}. \tag{13}$$

Having thus obtained a heuristic correctional scaling factor, we replace the error estimator by

$$\Delta_{\mathsf{sc}}^{\mathsf{pr}}(u) = \frac{\Delta_{n_t}^{\mathsf{pr}}(u)}{C_{\mathsf{sc}}}.$$

Fig. 3 For some of the random controls in Table 1, these are the cost function values $\hat{J}^h_{1/2}(u)$ and $\hat{J}^{h\ell}_{1/2}(u)$. Unvisible due to size, the reduced-order cost function plots are accompanied by an error bar of size $\Delta^J_i(u)$ for $i = 1, 2$

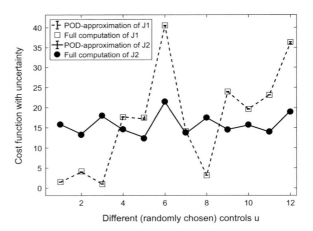

At this point, we would like to give two examples for the effect of the presented error estimation without going into more detail of the concrete numerical setup. All results which are presented here were evaluated for the model described in Sect. 4.

In Table 1, we illustrate the result of the scaling procedure for randomly chosen controls from within the admissible box $\mathcal{U}_{\mathrm{ad}}$ and the first cost function $\hat{J}^h_1$. It can be observed that the mean and minimum efficiency of the estimator is improved when moving from the unscaled to the scaled state estimator. However, we lose mathematical rigorosity for the estimator, i.e., it may happen that the estimator is smaller than the true error, which can be observed in the fact that the maximum efficiency value for the scaled estimator exceeds the value 1.

In Fig. 3, we present some function values for the reduced-order and high-fidelity function evaluations. The error margin indicated by the estimator Δ^J is so small that it is not visible. When zooming in, we can see that the high-fidelity cost function lies roughly within the error bar. It is therefore apparent that the performance of the estimator is satisfactory for the purposes at hand. Therefore, we will utilize this estimator instead of $\Delta^{\mathrm{pr}}_{n_t}$ and define the according cost function estimator which will replace Δ^J:

$$\Delta^J_{sc} : \mathcal{U} \to \mathbb{R}^2, \quad \Delta^J_{sc,i}(u) := \sqrt{2\Delta^{\mathrm{pr}}_{sc}(u)\,\hat{J}^{h\ell}_i(u)} + \tfrac{1}{2}\Delta^{\mathrm{pr}}_{sc}(u) \tag{14}$$

U-Local Reduced Bases: A Numerical Algorithm

In this section, we will combine the above results with the inexact subdivision algorithm presented in the section "Inexact Function Values" in order to efficiently and globally solve PDE-constrained MOCPs with set-oriented techniques. To achieve the desired efficiency, the error estimates will be tightened by a heuristic factor and then concepts from a u-local reduced basis method will be adapted to the multiobjective setting. Due to the nature of the subdivision algorithm, large parts of the decision space are discarded in early iterations such that it is favorable to begin with a relatively coarse approximation by ROMs and then later add models in regions that are close to the Pareto set.

When using a reduced-order model to approximate the function values $\hat{j}^{h\ell}$, we have to ensure that the error estimator $\Delta_{sc}^{J}(u)$ is less than a prescribed bound $\Delta_{sc,max}^{J} \in \mathbb{R}^{k}$ everywhere in the parameter domain. If we want to achieve this goal with a single ROM, the model may become high-dimensional in order to satisfy the error bounds everywhere and hence, inefficient. We therefore adopt ideas from u-local reduced basis methods [8, 13] and construct a library $\mathscr{L}$ of locally valid models during the subdivision procedure. This library can grow or shrink in every iteration of the algorithm.

In each subdivision step, all sample points (the indices of which are contained in the set $\mathscr{N}$) are evaluated using the *closest* ROM in $\mathscr{L}$, where the distance is defined via the Euclidean distance between the control u and the reference control u_{ref}^{j} at which the jth ROM was created. In the beginning, all points are denoted as *insufficiently approximated*:

$$\mathscr{I} = \left\{ i \in \mathscr{N} \mid \Delta_{sc}^{J}(u^{i}) \not\leq \Delta_{sc,max}^{J} \right\},$$

meaning that they have not been approximated well enough using a ROM. Then all sample points are evaluated using the available ROMs in $\mathscr{L}$ and the objective values $\hat{J}^{\ell}$ as well as the error estimates according to (11) are computed. All points with a satisfactory error estimate are eliminated from $\mathscr{I}$. Since the remaining points violate the desired error bound $\Delta_{sc,max}^{J} \in \mathbb{R}^{k}$, we evaluate the full model and add a ROM to the library $\mathscr{L}$. This is done in a *greedy* way (see also [11]), i.e., we add the ROM at the point with the maximum error. Here, we choose the basis size ℓ in such a way that the error estimator satisfies $\Delta_{sc}^{J}(u_{ref}) < \sigma \Delta_{sc,max}^{J}$ for some $\sigma \in [0, 1]$. Choosing $\sigma < 1$ results in an increased basis size ℓ but on the other hand very likely increases the region within $\mathcal{U}_{ad}$ where the ROM is valid.

The above steps are repeated until all points are approximated sufficiently accurately and consequently, the set $\mathscr{I}$ is empty. Finally, all ROMs are removed from $\mathscr{L}$ which have not been used. This is done in order to keep the number of locally valid ROMs at an acceptable number. Moreover, ROMs belonging to regions in the parameter domain which have been identified as dominated will not be required any further. The procedure for evaluating the sample points is summarized in Algorithm 3.

We want to emphasize that the approach presented here is only a first step toward using local reduced bases within multiobjective optimization. We expect that the efficiency can be further increased by implementing more sophisticated rules for clustering the points than using the Euclidean distance. Moreover, we expect that making use of online enrichment (see, e.g., [13]) or a combination of different bases will be beneficial for the overall performance.

4 Numerical Results

In this section, we apply Algorithm 4 to Problem ($\hat{\mathbf{P}}^h$) and compare the results to an FEM-based solution obtained by applying the standard subdivision algorithm (Algorithm 1).

Algorithm 3 (Greedy u-local reduced basis approach)

Require: $\Delta_{\mathsf{sc,max}}^J \in \mathbb{R}^k$, Library of locally valid ROMS $\mathscr{L}$, Scaling factor C_{sc}, local error scaling factor σ, set of sample points $\mathscr{N} \subset \mathcal{U}_{\mathsf{ad}}$;

1: Consider all sample points as *insufficiently approximated*, i.e. $\mathscr{I} = \mathscr{N}$;
2: **while** $\mathscr{I} \neq \emptyset$ **do**
3:    **for** $i = 1, \ldots, |\mathscr{I}|$ **do**
4:       Identify the *closest* ROM with respect to the 2-norm:

$$\hat{\imath} = \arg\min_{j \in \{1, \ldots, |\mathscr{L}|\}} |u^i - u_{\mathsf{ref}}^j|_2.$$

5:       Compute $\hat{J}^{h\ell}(u^i)$ using ROM $\hat{\imath}$;
6:       Evaluate the error $\Delta_{\mathsf{sc}}^J(u^i)$ for ROM $\hat{\imath}$ using (14);
7:       **if** $\Delta_{\mathsf{sc}}^J(u^i) \leq \Delta_{\mathsf{sc,max}}^J$ **then**
8:          Accept $\hat{J}^{h\ell}(u^i)$ as sufficiently accurate;
9:          Remove i from the set $\mathscr{I}$;
10:    Identify the sample point with the largest error:

$$i_{\mathsf{max}} = \arg\max_{s \in \mathscr{I}} \Delta_{\mathsf{sc}}^J(u^s)$$

11:    Add ROM to library $\mathscr{L}$ with $u_{\mathsf{ref}} = u^{i_{\mathsf{max}}}$ and ℓ such that $\Delta_{\mathsf{sc}}^J(u_{\mathsf{ref}}) < \sigma \Delta_{\mathsf{sc,max}}^J$;
12: Remove all ROMs from $\mathscr{L}$ that have not been used;

Problem Specification and General Setup

As the domain, we consider the unit square $\Omega = (0, 1)^2$ and the time interval $[0, T] = [0, 1]$. The desired states are given by

$$y_{d,1}(x) = \begin{cases} 0.5, & x_2 \leq 0.5, \\ 0.3, & x_2 > 0.5, \end{cases} \quad y_{d,2}(x) = \begin{cases} -0.5, & x_1 \leq 0.5, \\ 0.5, & x_1 > 0.5, \end{cases}$$

such that there are both conflicting and non-conflicting areas in the domain. The subdomains are given by

$$\Omega_1 = [0, 0.5] \times [0, 0.5], \quad \Omega_2 = [0, 0.5] \times (0.5, 1],$$
$$\Omega_3 = (0.5, 1] \times [0, 0.5], \quad \Omega_4 = (0.5, 1] \times (0.5, 1].$$

The initial condition is $y_0(x) = 0$ for all $x \in \Omega$, and we allow controls for the constraints $u_b = (1, 1, 1, 1)^\top$ and $u_a = -u_b$.

We chose linear finite elements on a triangular grid over Ω containing $n = 712$ degrees of freedom. The time interval is split into an equidistant grid of $n_t = 101$ points.

Algorithmic Results

Algorithm 4 (POD-based inexact subdivision algorithm)

Require: $\Delta^J_{sc,max} \in \mathbb{R}^k$, set of sample controls $\mathcal{U}_{sc}$;
1: Compute the heuristic factor C_{sc} according to (13);
2: Compute the ROMs corresponding to the sample controls $u \in \mathcal{U}_{sc}$ and store them in the library $\mathscr{L}$;
3: Call the Algorithm 2 wherein every selection step, the sample points are evaluated according to Algorithm 3;

The entire numerical scheme for the POD-based approach for solving ($\hat{\mathbf{P}}^h$) is summarized in Algorithm 4. The first step is to compute the heuristic factor C_{sc} using several ROMs at randomly distributed controls $u_{ref} \in \mathcal{U}_{sc}$ within the parameter domain. In each of these points, the factor between the true error and the error estimator is computed and we set C_{sc} as the mean value of these computations. Note that in our example, this factor was approximately 2.5 everywhere in the parameter domain such that this approach is justified.[1] The library $\mathscr{L}$ is initialized by constructing a ROM for each of these FEM solutions. The parameter σ which is used to determine the local basis size (cf. section "U-Local Reduced Bases: A Numerical Algorithm") has been set to 0.5 which yields basis sizes between 4 and 16 for the current setting.

Each box B is represented by an equidistant grid of two points in each direction, i.e., by 16 sample points in total. In the standard approach (Algorithm 1), one FEM evaluation is required for each evaluation of the objective function. The exact Pareto set of ($\hat{\mathbf{P}}^h$) is shown in Fig. 4a, where the boxes are colored according to the fourth component u_4. The corresponding Pareto front is given in Fig. 4c in green. The Pareto set for the same problem, obtained by Algorithm 4, is shown in Fig. 4b, the corresponding Pareto front in Fig. 4c in red. We observe a good agreement both between the Pareto sets and the Pareto fronts. The error bound $\Delta^J_{sc,max} = (0.025, 0.025, 0)^\top$ is satisfied as desired. It should be mentioned that the two Pareto sets are still visibly different. Due the fact that the Pareto front is very steep, the inexact non-dominance property (Definition 4) results in a significantly reduced number of dominated boxes

---

[1] In a scalar optimization problem, it would not be advisable to give up rigorosity in order to reduce the error by such a factor. However, when dealing with multiple objectives and 'steep' Pareto fronts (cf. Remark 4), this factor results in huge computational savings since the number of dominated boxes is drastically reduced in every step.

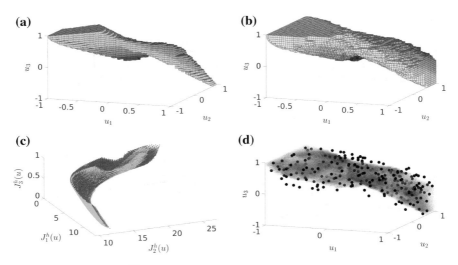

Fig. 4 **a** Pareto set of $(\hat{\mathbf{P}}^h)$ after 22 subdivision steps using a finite element discretization. Projection onto the first three components of u, and u_4 is visualized by the box coloring. **b** The Pareto set based on POD reduced-order models and the inexact sampling algorithm with $\Delta_{\mathrm{sc,max}}^J = (0.025, 0.025, 0)^\top$. **c** The corresponding Pareto fronts, where the FEM-based solution is depicted in green and the POD-based solution in red. The points are the images of the box centers. **d** Clustering of the sample points. Each of the colored patches has been assigned to one ROM, which are represented by black dots

in each subdivision step. When further decreasing the error bound $\Delta_{\mathrm{sc,max}}^J$, this results in ROMs that are only valid for very few sample points such that the number of FEM evaluations is not sufficiently reduced. This emphasizes the additional difficulties introduced by multiple objectives and justifies the use of a heuristic scaling factor. Furthermore, the error bound only influences the distance between the exact and the inexact the Pareto fronts. In order to also bound the error in the decision space, further assumptions on the objectives have to be made.

For the computation of the inexact Pareto set, one ROM evaluation is required for each evaluation of the objective function. In order to realize this with the prescribed accuracy, 444 local ROMs were created throughout Algorithm 4. This means that in sum over all subdivision steps, only 444 evaluations of the FEM model were required (cf. Fig. 5a). Compared to the standard approach, the number of FEM evaluations was reduced by a factor of more than 1000. The FEM evaluations are mainly performed during the first subdivision steps (cf. Fig. 5a), whereas later, almost the entire decision space can be approximated by the existing models. This is also illustrated in Fig. 5b, where we observe an exponential increase of the ratio of high-fidelity solutions between the FEM and the POD-based approach. Therefore, it might be interesting to investigate the benefit of an offline phase similar to classical reduced basis approaches. However, as has been mentioned before, this should ideally be

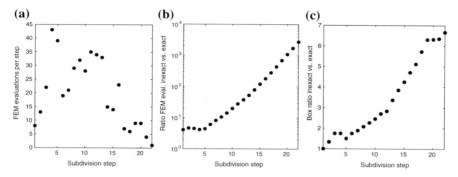

Fig. 5 **a** The number of FEM evaluations in each subdivision step in the POD-based inexact subdivision algorithm 4. **b** Ratio of the total number of FEM evaluations within the POD-based inexact subdivision algorithm 4 and the FEM-based exact subdivision algorithm 1 up to the current step. **c** Ratio of the respective number of boxes. This is directly influenced by $\Delta^J_{sc,max}$

done in such a way that not too much numerical effort is invested in parts of the decision space which are far apart from the Pareto set.

Due to the inexactness, the number of boxes is larger in the inexact computation, which is shown in Fig. 5c. Similar to the observations in [20], this effect becomes more severe during higher-order subdivision steps. All boxes in the vicinity of the exact Pareto set are not eliminated which results in an exponential increase in the number of boxes. This ratio is directly influenced by the error bound $\Delta^J_{sc,max}$. In other words, the combination of the inexact subdivision algorithm with POD-based reduced-order modeling leads to a dilemma: if we demand a very small error for the ROM, the number of boxes in the exact and the inexact solution will be very similar. On the other hand, this results in very accurate and therefore less efficient ROMs. Consequently, further research is required to address this issue and determine an optimal trade-off in terms of efficiency.

In Fig. 4d, the locations of the reference controls u_{ref} of the (remaining) ROMs are shown as black dots. The remaining points depict the sample points that were evaluated in the 22nd subdivision step, where the coloring depends on the ROM the point was assigned to. Due to the fact that the selection is being based on the Euclidean distance the size of all the patches is comparable. Since there is no formal reason to choose this specific way of assigning sample points to ROMs, this motivates the investigation of more advanced clustering approaches in order to further reduce the number of required ROMs.

5 Conclusion and Outlook

In this chapter, we have applied an inexact version of the gradient-free subdivision algorithm from [20] to a multiobjective optimal control problem with a semilinear parabolic state equation. The inexactness was in particular given by employing POD model-order reduction to the state equation in order to speed up cost function evaluations. Due to the existence of analytical error estimates for the resulting state error, we were able to present an a posteriori estimator for the cost function error. The maximal error for the reduced-order model is given a priori by the problem definition. In order to ensure the necessary accuracy, we employ a u-local basis strategy. The ROMs are constructed iteratively using a Greedy worst-first search. Our numerical results for a simple test problem confirm that the reduced-order approach is able to qualitatively approximate both the Pareto front and the Pareto set. The Pareto front is captured accurately according the prescribed error bounds on the cost function values. Furthermore, the error estimator for the state variable and the cost function values are coupled with a simple heuristic to compensate for a constant over-estimation factor.

It has to be mentioned that there is room for improvement to the presented algorithm. Due to the difference in the Definitions 3 (exact dominance) and 4 (inexact dominance), it is a natural consequence that the inexact algorithm can eliminate fewer boxes in each iteration. This makes it necessary to perform more function eliminations in the next iteration than for the exact algorithm. It may therefore be beneficial to develop alternative or additional criteria in order reduce the number of non-dominated boxes.

As far as the construction of reduced-order models is concerned, it has to be stated that the current strategy only follows the heuristic that controls which lie close to each other in a geometrical sense may share similar state and cost function behavior. This straightforward approach may result in an inefficiently large number of reduced-order models since there may be more distant controls which are still sufficiently well approximated by the given reduced model. As a result, more sophisticated clustering techniques in the decision space need to be developed which result in fewer, more cleverly spread reduced models for the state equation.

The gradient-based version of the subdivision algorithm has also been extended to inexactness in [20]. In order to combine model-order reduction and in particular error estimation with this method, we would have to estimate the error between the gradients $\nabla \hat{J}_i^{h\ell}(u)$ and the reduced-order gradients $\nabla \hat{J}_i^{h\ell}(u)$ by using an a posteriori estimator for the *adjoint equation* to the system (1).

Acknowledgements This work is supported by the Priority Programme SPP 1962 *Non-smooth and Complementarity-based Distributed Parameter Systems* of the German Research Foundation (DFG) and by the project *Hybrides Planungsverfahren zur energieeffizienten Wärme- und Stromversorgung von städtischen Verteilnetzen* funded by the German Ministry for Economic Affairs and Energy.

References

1. Albrecht, F., Haasdonk, B., Kaulmann, S., Ohlberger, M.: The localized reduced basis multi-scale method. In: Proceedings of Algorithmy 2012, pp. 393–403 (2012)
2. Albunni, M.N., Rischmuller, V., Fritzsche, T., Lohmann, B.: Multiobjective optimization of the design of nonlinear electromagnetic systems using parametric reduced order models. IEEE Trans. Magn. **45**(3), 1474–1477 (2009)
3. Banholzer, S., Beermann, D., Volkwein, S.: POD-based bicriterial optimal control by the reference point method. IFAC-PapersOnLine **49**, 210–215 (2016)
4. Banholzer, S., Beermann, D., Volkwein, S.: POD-based error control for reduced-order bicriterial PDE-constrained optimization. Annu. Rev. Control (2017). (To appear)
5. Chaturantabut, S., Sorensen, D.C.: Nonlinear model reduction via discrete empirical interpolation. SIAM J. Sci. Comput. **32**(5), 2737–2764 (2010)
6. Coello Coello, C.A., Lamont, G.B., Van Veldhuizen, D.A.: Evolutionary Algorithms for Solving Multi-Objective Problems, vol. 2. Springer Science & Business Media (2007)
7. Dellnitz, M., Schütze, O., Hestermeyer, T.: Covering Pareto sets by multilevel subdivision techniques. J. Optim. Theory Appl. **124**(1), 113–136 (2005)
8. Dihlmann, M., Kaulmann, S., Haasdonk, B.: Online reduced basis construction procedure for model reduction of parametrized evolution systems. In: IFAC Proceedings Volumes, vol. 45, no. 2, pp. 112–117 (2012)
9. Drohmann, M., Haasdonk, B., Ohlberger, M.: Reduced basis approximation for nonlinear parametrized evolution equations based on empirical operator interpolation. SIAM J. Sci. Comput. **34**(2), A937–A969 (2012)
10. Ehrgott, M.: Multicriteria Optimization, 2nd edn. Springer, Berlin Heidelberg New York (2005)
11. Grepl, M.A., Maday, Y., Nguyen, N.C., Patera, A.T.: Efficient reduced-basis treatment of non-affine and nonlinear partial differential equations. ESAIM. Math. Model. Numer. Anal. **41**, 575–605 (2007)
12. Gubisch, M., Volkwein, S.: Proper orthogonal decomposition for linear-quadratic optimal control. In Ohlberger, M., Benner, P., Cohen, A., Willcox, K. (eds.) Model Reduction and Approximation: Theory and Algorithms, pp. 5–66. SIAM, Philadelphia, PA (2017)
13. Haasdonk, B., Dihlmann, M., Ohlberger, M.: A training set and multiple bases generation approach for parameterized model reduction based on adaptive grids in parameter space. Mathe. Comput. Model. Dyn. Syst. **17**(4), 423–442 (2011)
14. Hillermeier, C.: Nonlinear Multiobjective Optimization: A Generalized Homotopy Approach. Birkhäuser (2001)
15. Holmes, P., Lumley, J.L., Berkooz, G., Rowley, C.W.: Turbulence, Coherent Structures, Dynamical Systems and Symmetry. Cambridge Monographs on Mechanics, 2nd edn. Cambridge University Press, Cambridge (2012)
16. L. Iapichino, S. Trenz, and S. Volkwein. Reduced-order multiobjective optimal control of semilinear parabolic problems. In: Numerical Mathematics and Advanced Applications (ENUMATH 2015). Lecture Notes in Computational Science and Engineering, vol. 112, pp. 389–397. Springer Switzerland (2016)
17. Iapichino, L., Ulbrich, S., Volkwein, S.: Multiobjective PDE-constrained optimization using the reduced-basis method. Adv. Comput. Math. (2017). (To appear)
18. Miettinen, K.: Nonlinear Multiobjective Optimization. Springer Science & Business Media (2012)

19. Ohlberger, M., Schindler, F.: Error control for the localized reduced basis multiscale method with adaptive on-line enrichment. SIAM J. Sci. Comput. **37**(6), 2865–2895 (2015)
20. Peitz, S., Dellnitz, M.: Gradient-based multiobjective optimization with uncertainties. In: NEO 2016 Proceedings. Springer (2017). arXiv:1612.03815. (to appear)
21. Peitz, S., Ober-Blöbaum, S., Dellnitz, M.: Multiobjective optimal control methods for fluid flow using model order reduction (2015). arXiv:1510.05819
22. Queipo, N.V., Haftka, R.T., Shyy, W., Goel, T., Vaidyanathan, R., Tucker, K.P.: Surrogate-based analysis and optimization. Prog. Aerosp. Sci. **41**, 1–28 (2005)
23. Rogg, S., Trenz, S., Volkwein, S.: Trust-region POD using a-posteriori error estimation for semi-linear parabolic optimal control problems. Konstanzer Schriften in Mathematik, 359 (2017). https://kops.uni-konstanz.de/handle/123456789/38240
24. Schilders, W.H.A., van der Vorst, H.A., Rommes, J.: Model Order Reduction. Springer, Berlin Heidelberg (2008)
25. Schütze, O.: A new data structure for the nondominance problem in multi-objective optimization. In: International Conference on Evolutionary Multi-Criterion Optimization, pp. 509–518. Springer Berlin Heidelberg (2003)
26. Schütze, O., Witting, K., Ober-Blöbaum, S., Dellnitz, M.: Set Oriented Methods for the Numerical Treatment of Multiobjective Optimization Problems. In: Tantar, E., Tantar, A.-A., Bouvry, P., Del Moral, P., Legrand, P., Coello Coello, C.A., Schütze, O. (eds.) EVOLVE-A Bridge between Probability, Set Oriented Numerics and Evolutionary Computation. Studies in Computational Intelligence, vol. 447, pp. 187–219. Springer Berlin Heidelberg (2013)
27. Singler, J.R.: New POD error expressions, error bounds, and asymptotic results for reduced order models of parabolic PDEs. SIAM J. Numer. Anal. **52**, 852–876 (2014)
28. Tröltzsch, F.: Optimal Control of Partial Differential Equations. Graduate Studies in Mathematics, vol. 112. American Mathematical Society (2010)

Sequential Reduced-Order Modeling for Time-Dependent Optimization Problems with Initial Value Controls

Matthias Heinkenschloss and Dörte Jando

Abstract This paper introduces an efficient sequential application of reduced order models (ROMs) to solve linear quadratic optimal control problems with initial value controls. The numerical solution of such a problem requires Hessian-times-vector multiplications, each of which requires solving a linearized state equation with initial value given by the vector and solving a second-order adjoint equation. Projection-based ROMs are applied to these differential equations to generate a Hessian approximation. However, in general, no fixed ROM well-approximates the application of the Hessian to all possible vectors of initial data. To improve a basic ROM, *Heinkenschloss and Jando: Reduced-Order Modeling for Time-Dependent Optimization Problems with Initial Value Controls (SIAM Journal on Scientific Computing, 40(1), A22–A51, 2018*, https://doi.org/10.1137/16M1109084*)* introduce an augmentation of the basic ROM by the right-hand side of the optimality system. This augmented ROM substantially improves the accuracy of the computed control, but this accuracy may still not be enough. The proposed sequential application of the augmented ROM can compute an approximate control with the same accuracy as the one obtained using only the expensive full-order model, but at a fraction of the cost.

Keywords Optimal control · Reduced-order models · Hessian approximation Initial-condition problems · Sequential reduced-order models

M. Heinkenschloss
Department of Computational and Applied Mathematics, Rice University,
MS-134, 6100 Main Street, Houston, TX 77005-1892, USA
e-mail: heinken@rice.edu

D. Jando (✉)
Interdisciplinary Center for Scientific Computing (IWR), Heidelberg University,
Im Neuenheimer Feld 205, 69120 Heidelberg, Germany
e-mail: doerte.jando@iwr.uni-heidelberg.de

© Springer International Publishing AG, part of Springer Nature 2018
W. Keiper et al. (eds.), *Reduced-Order Modeling (ROM) for Simulation and Optimization*,
https://doi.org/10.1007/978-3-319-75319-5_4

1 Introduction

We present an efficient approach for the sequential application of reduced-order models (ROMs) introduced in [8] to solve a class of large-scale linear quadratic optimal control problems with initial value controls. Such problems arise, e.g., from semi-discretizations of optimal control problems governed by partial differential equations (PDEs) in the context of source inversion problems or as subproblems in multiple shooting formulations of optimal control problems. The solution of a linear quadratic optimal control problem with initial value controls is characterized by a large linear optimality system with system matrix given by the Hessian of the optimal control problem. The computation of a Hessian–vector product requires the solution of the linearized state equation with initial value given by the vector to which the Hessian is applied to, followed by the solution of the second-order adjoint equation. In our recent work [8], projection-based ROMs of these two linear differential equations are used to generate the Hessian approximation. The main novelty of [8] is the augmentation of a basic ROM by the right-hand side vector of the optimality system. Although the size of the ROM increases only by one, it was proven in [8] that this new augmented ROM produces substantially better approximations than the basic ROM. However, while the augmented ROM approach of [8] provides a substantial improvement over the basic ROM, the resulting approximation of the initial value control may not be sufficiently accurate compared to the approximation computed using the expensive full-order model. This paper extends [8] by applying the ROM approach sequentially. Each step in the sequential application uses a small-sized ROM that works on a different subspace. Projections are used to isolate these subspaces. The augmented ROM approach of [8] provides even greater benefits in this new sequential ROM application. A numerical example shows that this new sequential ROM application can compute an approximate solution of the optimal control problem with the same accuracy as the solution computed using the expensive full-order model, but at a fraction of the cost (about 40% in our example).

Given symmetric positive definite matrices $\mathbf{M}, \mathbf{R} \in \mathbb{R}^{n \times n}$, symmetric positive semidefinite matrices $\mathbf{Q}, \mathbf{Q}_T \in \mathbb{R}^{n \times n}$, a matrix $\mathbf{A} \in \mathbb{R}^{n \times n}$, functions $\mathbf{c}, \mathbf{f} \in L^2(0, T; \mathbb{R}^n)$, and a vector $\mathbf{c}_T \in \mathbb{R}^n$, we consider the linear quadratic optimal control problem

$$\min_{\mathbf{s} \in \mathbb{R}^n} \ J(\mathbf{s}) \tag{1a}$$

with objective function

$$J(\mathbf{s}) = \int_0^T \tfrac{1}{2} \mathbf{y}(t)^T \mathbf{Q} \mathbf{y}(t) + \mathbf{c}(t)^T \mathbf{y}(t) dt + \tfrac{1}{2} \mathbf{y}(T)^T \mathbf{Q}_T \mathbf{y}(T) + \mathbf{c}_T^T \mathbf{y}(T) + \tfrac{1}{2} \mathbf{s}^T \mathbf{R} \mathbf{s}, \tag{1b}$$

where for given $\mathbf{s}$ the state $\mathbf{y} \in H^1(0, T; \mathbb{R}^n)$ solves

$$\mathbf{M}\frac{d}{dt}\mathbf{y}(t) + \mathbf{A}\mathbf{y}(t) = \mathbf{f}(t), \qquad\qquad t \in (0, T) \qquad (1c)$$

$$\mathbf{M}\mathbf{y}(0) = \mathbf{M}\mathbf{s}. \qquad\qquad (1d)$$

As we have already mentioned, problems of the type (1) arise, e.g., from semi-discretizations of optimal control problems governed by partial differential equations (PDEs) in the context of source inversion problems (see, e.g., Akcelik et al. [1], Bashir et al. [2]), as subproblems in data assimilation (see, e.g., Blum et al. [4, Sect. 6], Rao and Sandu [13], Daescu and Navon [6]), or as subproblems in multiple shooting formulations of optimal control problems (see, e.g., Hesse and Kanschat [9] or Carraro and Geiger [5]). The papers present solution approaches for (1). We refer to [8] for a complete review of the solution methods and comparison with the one presented in this paper.

The vast majority of approaches to apply ROMs for the solution of large-scale optimization problems are for problems with fixed initial data and control in the right-hand side. See, e.g., the overview articles by Benner et al. [3], Gubisch and Volkwein [7], or Sachs and Volkwein [14]. A major difficulty in applying ROMs to the solution of (1) arises from the fact that the initial data $\mathbf{s}$ are variable and can range over all of $\mathbb{R}^n$. This paper is one of a few where ROMs are systematically applied for problems where the initial data are variable.

Given a symmetric positive-definite matrix $\mathbf{G} \in \mathbb{R}^{n \times n}$, we endow $\mathbb{R}^n$ with the weighted inner product

$$\langle \mathbf{s}_1, \mathbf{s}_2 \rangle = \mathbf{s}_1^T \mathbf{G} \mathbf{s}_2 \qquad\qquad (2)$$

and corresponding norm. Although our augmented ROM approach in [8] applies to (1), the sequential approach introduced in this paper currently requires that

$$\mathbf{A}^T = \mathbf{A}, \quad \mathbf{M}^T = \mathbf{M}, \quad \mathbf{Q} = \mathbf{Q}_T = \mathbf{M}, \quad \mathbf{R} = \beta\mathbf{G}, \;\; \beta > 0, \quad \mathbf{G} = \mathbf{M}. \qquad (3)$$

This paper is organized as follows. The next section reviews the optimality system for (1) and the Hessian $\nabla^2 J$ computation. This section is essentially identically to [8, Sect. 2.1] but needs to be repeated to introduce background and notation for the sequential approach. Section 3 first outlines the sequential approach we take in this paper. Then, it reviews the Hessian approximation with the basic and the augmented ROM of [8], which in modified form are the basic building blocks of the proposed sequential approach. Section 4 uses the assumptions (3) to transform (1) into n-decoupled scalar optimal control problems. This was already done in [8] to gain insight into the performance of the basic ROM and the augmented ROM. Here we review these basic results but also extract important insights about the sequential application of these ROM approaches. Section 5 specifies our approach for the sequential application of both the basic ROM and the augmented ROM. The performance of our new sequential ROM application is demonstrated in Sect. 6 on an optimal control problem governed by a parabolic PDE. The sequential application of the augmented ROM outperforms the sequential application of the basic ROM. In particular, for our example, the sequential application of the augmented ROM computes an approximate solution of the same accuracy as the solution computed using the expensive full-order model, but the computational cost of the augmented ROM approach is about 40% of the computational cost of using only the full-order model.

2 Optimality Conditions and Hessian Computations

This section is essentially identical to [8, Sect. 2.1]. It reviews computation of the gradient and the Hessian of the objective function (1b). This review is necessary to introduce background and notation for the following sections.

Since the function (1b) is quadratic and strictly convex, (1) has a unique solution $\mathbf{s}^*$ and this solution is characterized by the condition $\nabla J(\mathbf{s}^*) = \mathbf{0}$. Since J is quadratic, its Hessian is independent of $\mathbf{s}$. The gradient and Hessian computation does not require assumptions (3), and these are not applied in this section.

The gradient of (1b) with respect to the inner product (2) can be computed via the adjoint equation approach (see, e.g., Hinze et al. [10, Chap. 1], or Liberzon [12]) and is given by

$$\nabla J(\mathbf{s}) = \mathbf{G}^{-1}\mathbf{M}\mathbf{p}(0) + \mathbf{G}^{-1}\mathbf{R}\mathbf{s} \in \mathbb{R}^n, \tag{4}$$

where $\mathbf{p} \in H^1(0, T; \mathbb{R}^n)$ is the solution of the adjoint equation

$$-\mathbf{M}\frac{d}{dt}\mathbf{p}(t) + \mathbf{A}^T\mathbf{p}(t) = \mathbf{Q}\,\mathbf{y}(t) + \mathbf{c}(t), \qquad t \in (0, T), \tag{5a}$$

$$\mathbf{M}\mathbf{p}(T) = \mathbf{Q}_T\mathbf{y}(T) + \mathbf{c}_T. \tag{5b}$$

The Hessian of J is independent of $\mathbf{s}$ and is given by

$$\nabla^2 J = \mathbf{H} + \mathbf{G}^{-1}\mathbf{R},$$

where the application of $\mathbf{H}$ to a vector $\mathbf{v} \in \mathbb{R}^n$ is given by

$$\mathbf{H}\mathbf{v} = \mathbf{G}^{-1}\mathbf{M}\mathbf{q}(0), \tag{6}$$

and $\mathbf{q} \in H^1(0, T; \mathbb{R}^n)$ is obtained by first solving

$$\mathbf{M}\frac{d}{dt}\mathbf{z}(t) + \mathbf{A}\mathbf{z}(t) = \mathbf{0}, \qquad t \in (0, T), \tag{7a}$$

$$\mathbf{M}\mathbf{z}(0) = \mathbf{M}\mathbf{v}, \tag{7b}$$

and then

$$-\mathbf{M}\frac{d}{dt}\mathbf{q}(t) + \mathbf{A}^T\mathbf{q}(t) = \mathbf{Q}\,\mathbf{z}(t), \qquad t \in (0, T), \tag{8a}$$

$$\mathbf{M}\mathbf{q}(T) = \mathbf{Q}_T\mathbf{z}(T). \tag{8b}$$

It is easy to verify (see [8]) that $\mathbf{H}$ is self-adjoint with respect to the inner product (2), i.e., satisfies $\mathbf{H} = \mathbf{H}^* = \mathbf{G}^{-1}\mathbf{H}^T\mathbf{G}$ and is positive semidefinite with respect to the inner product (2). Thus, the Hessian $\nabla^2 J$ is self-adjoint and positive definite with respect to the inner product (2).

3 Overview of Our Solution Approach

Iteration

Given an approximation $\mathbf{s}^{(0)}$ of the solution $\mathbf{s}^*$ of (1), we want to compute a correction $\Delta\mathbf{s}$ such that $\mathbf{s}_* = \mathbf{s}^{(0)} + \Delta\mathbf{s}$. Since $\mathbf{0} = \nabla J(\mathbf{s}^{(0)} + \Delta\mathbf{s}) = \nabla J(\mathbf{s}^{(0)}) + \nabla^2 J \,\Delta\mathbf{s}$, the correction is the solution of

$$(\mathbf{H} + \mathbf{G}^{-1}\mathbf{R}) \,\Delta\mathbf{s} = -\nabla J(\mathbf{s}^{(0)}). \tag{9}$$

The system (9) can be solved using the conjugate gradient (CG) method. Each iteration of the CG method requires the computation of a Hessian-times-vector product; i.e., each iteration requires the solution of (7) and (8).

To reduce the computational cost of these two simulations, we will use ROMs. Applying a ROM to (7) and (8), we arrive at an approximation $\widehat{\mathbf{H}}$ of $\mathbf{H}$. Thus we solve

$$(\widehat{\mathbf{H}} + \mathbf{G}^{-1}\mathbf{R}) \,\Delta\mathbf{s} = -\nabla J(\mathbf{s}^{(0)}) \tag{10}$$

instead of (9). Note that ROMs are only applied to approximate the Hessian, but not to compute $\nabla J(\mathbf{s}^{(0)})$. In [8], we constructed (10) and used the resulting $\mathbf{s}^{(0)} + \Delta\mathbf{s}$ to approximate $\mathbf{s}^*$. This may not be sufficient.

In this paper, we use an iterative approach. In the ℓth step, given $\mathbf{s}^{(\ell)}$, we compute a ROM, which depends on $\mathbf{s}^{(\ell)}$ and construct an approximation $\widehat{\mathbf{H}}^{(\ell)}$ of $\mathbf{H}$. Then we solve a version of

$$(\widehat{\mathbf{H}}^{(\ell)} + \mathbf{G}^{-1}\mathbf{R}) \,\Delta\mathbf{s} = -\nabla J(\mathbf{s}^{(\ell)}), \tag{11}$$

set $\mathbf{s}^{(\ell+1)} = \mathbf{s}^{(\ell)} + \Delta\mathbf{s}$ and repeat the process if necessary. The structure of $\widehat{\mathbf{H}}^{(\ell)}$ is identical to that in [8], which we will review in the next subsection, except that the ROM is adjusted in every iteration. In Sects. 4 and 5, we will specify this adjustment and also what we mean by 'solve a version of (11).' Again it is important to note that for each $\mathbf{s}^{(\ell)}$, we compute the exact gradient $\nabla J(\mathbf{s}^{(\ell)})$ at the expense of solving the full-order systems (1c,d) and (5). Thus, we can use $\|\nabla J(\mathbf{s}^{(\ell)})\|$ to monitor the error between $\mathbf{s}^{(\ell)}$ and $\mathbf{s}^*$, just as we would do when we solve the full-order problem directly. We only use ROMs to approximate the Hessian to expedite the solution approach.

The structure of the Hessian approximations $\widehat{\mathbf{H}}^{(\ell)}$ is identical to that in [8]. In [8], we allowed nonsymmetric $\mathbf{A}$. In the following, we will review the Hessian approximation of [8] for symmetric $\mathbf{A}$.

Reduced-Order Model Hessian Approximation

Assume $\mathbf{A}^T = \mathbf{A}$. Let the matrix for model-order reduction be $\mathbf{V} \in \mathbb{R}^{n \times k}$, $k < n$, such that $\mathbf{V}^T \mathbf{M} \mathbf{V}$ is invertible (we will construct $\mathbf{V}$ with $\mathbf{V}^T \mathbf{M} \mathbf{V} = \mathbf{I}_k$). The ROM approximations of the solutions to (7) and (8) are given by $\mathbf{V}\widehat{\mathbf{z}}$ and $\mathbf{V}\widehat{\mathbf{q}}$, respectively, where $\widehat{\mathbf{z}}$, $\widehat{\mathbf{q}}$ solve

$$\mathbf{V}^T \mathbf{M} \mathbf{V} \frac{d}{dt} \widehat{\mathbf{z}}(t) + \mathbf{V}^T \mathbf{A} \mathbf{V} \, \widehat{\mathbf{z}}(t) = \mathbf{0}, \qquad\qquad t \in (0, T), \qquad (12a)$$

$$\mathbf{V}^T \mathbf{M} \mathbf{V} \widehat{\mathbf{z}}(0) = \mathbf{V}^T \mathbf{M} \mathbf{s} \qquad\qquad (12b)$$

and

$$-\mathbf{V}^T \mathbf{M} \mathbf{V} \frac{d}{dt} \widehat{\mathbf{q}}(t) + \mathbf{V}^T \mathbf{A}^T \mathbf{V} \, \widehat{\mathbf{q}}(t) = \mathbf{V}^T \mathbf{Q} \mathbf{V} \, \widehat{\mathbf{z}}(t), \qquad t \in (0, T), \qquad (13a)$$

$$\mathbf{V}^T \mathbf{M} \mathbf{V} \widehat{\mathbf{q}}(T) = \mathbf{V}^T \mathbf{Q}_T \mathbf{V} \, \widehat{\mathbf{z}}(T). \qquad\qquad (13b)$$

The ROM approximation of (6) is given by

$$\widehat{\mathbf{H}} \mathbf{s} = \mathbf{G}^{-1} \mathbf{M} \mathbf{V} \widehat{\mathbf{q}}(0). \qquad\qquad (14)$$

We showed in [8] that $\widehat{\mathbf{H}}$ is self-adjoint and positive semidefinite with respect to the inner product (2).

Our Hessian approximations $\widehat{\mathbf{H}}^{(\ell)}$ are of the form (14) but with different ROM matrices $\mathbf{V}$. While the ROM matrices $\mathbf{V}$ will be adjusted during the iteration, their construction will be based on that in [8], which we will summarize next. While we allowed nonsymmetric $\mathbf{A}$ in [8], we continue to assume $\mathbf{A}^T = \mathbf{A}$.

Basic ROM and Augmented ROM

Assume $\mathbf{A}^T = \mathbf{A}$. Let $\lambda_i \in \mathbb{R}$ be the generalized eigenvalues of $(\mathbf{A}, \mathbf{M})$, sorted in ascending order, i.e.,

$$\lambda_1 \leqslant \cdots \leqslant \lambda_n, \qquad\qquad (15a)$$

and let $\mathbf{V}_n \in \mathbb{R}^{n \times n}$ be the matrix whose columns are the $\mathbf{M}$-orthonormal eigenvectors. For $\mathbf{\Lambda} = \mathrm{diag}(\lambda_1, \ldots, \lambda_n) \in \mathbb{R}^{n \times n}$, we have

$$\mathbf{A} \mathbf{V}_n = \mathbf{M} \mathbf{V}_n \mathbf{\Lambda}, \quad \mathbf{V}_n^T \mathbf{M} \mathbf{V}_n = \mathbf{I}. \qquad\qquad (15b)$$

Basic Reduced-Order Model

The initial (step $\ell = 0$) basic ROM $\mathbf{V} \in \mathbb{R}^{n \times k}$ is obtained by taking the first k columns of $\mathbf{V}_n$. This ROM includes the k eigenvalues corresponding to the most important solution components in (7) and (8). Under the assumptions (3), this ROM also leads to $\widehat{\mathbf{H}}$ that is the best rank k approximation of $\mathbf{H}$ (see [8] and subsection "Sequential Hessian Approximation Using the Basic ROM" in Sect. 4 for details). However, if we want to solve $(\mathbf{H} + \mathbf{G}^{-1}\mathbf{R})\mathbf{s} = -\mathbf{g}$ then approximating $\mathbf{H}$ by its best rank k approximation is not necessarily sufficient. The right-hand side $\mathbf{g}$ needs to be captured sufficiently well.

Augmented Reduced-Order Model

To capture the right-hand side $\mathbf{g}$ of the linear system $(\mathbf{H} + \mathbf{G}^{-1}\mathbf{R})\mathbf{s} = -\mathbf{g}$, we proposed in [8] to augment the basic ROM $\mathbf{V} \in \mathbb{R}^{n \times k}$ as follows:

$$\widehat{\mathbf{V}} = [\mathbf{V}, \mathbf{v}] \in \mathbb{R}^{n \times (k+1)}, \tag{16a}$$

where

$$\mathbf{v} = (\mathbf{g} - \mathbf{V}\, \mathbf{V}^T \mathbf{Mg})/\|\mathbf{g} - \mathbf{V}\, \mathbf{V}^T \mathbf{Mg}\|_{\mathbf{M}} \tag{16b}$$

is the $\mathbf{M}$-orthogonal projection of $\mathbf{g}$ onto the complement of $\mathbf{V}$. By construction $\text{range}(\widehat{\mathbf{V}}) = \text{range}([\mathbf{V}, \mathbf{g}])$ and $\widehat{\mathbf{V}}$ is $\mathbf{M}$-orthonormal; i.e., $\widehat{\mathbf{V}}^T \mathbf{M} \widehat{\mathbf{V}} = \mathbf{I}$.

4 Sequential Basic ROM Hessians for Decoupled Scalar Problems

Under the assumptions (3), a change of variables can be applied to decouple the problem (1) into n real scalar problems. This is used in [8] to analyze the ROM Hessian approximations presented in Sect. 3. We will review the main results here and also gain insight into how to construct reduced-order Hessian approximations for the sequential approach sketched previously in subsection "Reduced-Order Model Hessian Approximation".

We assume that (3) holds. As introduced previously in subsection "Basic ROM and Augmented ROM", $\lambda_i \in \mathbb{R}$ are the generalized eigenvalues of $(\mathbf{A}, \mathbf{M})$ and $\mathbf{V}_n \in \mathbb{R}^{n \times n}$ is the matrix whose columns are the $\mathbf{M}$-orthonormal eigenvectors, see (15).

Optimal Control Problem

As already done in [8], we transform the optimal control problem (1) into a problem with initial data

$$\widetilde{\mathbf{s}} \equiv \mathbf{V}_n^T \mathbf{M} \mathbf{s} \in \mathbb{R}^n$$

and with states $\widetilde{\mathbf{y}}$ defined through the identify[1]

$$\mathbf{y}(t) = \mathbf{V}_n \widetilde{\mathbf{y}}(t).$$

We define

$$\widetilde{\mathbf{f}}(t) = \mathbf{V}_n^T \mathbf{f}(t), \qquad \widetilde{\mathbf{c}}(t) = \mathbf{V}_n^T \mathbf{c}(t), \qquad \widetilde{\mathbf{c}}_T = \mathbf{V}_n^T \mathbf{c}_T,$$

and use (3) and (15b) to transform (1) into

$$\min_{\widetilde{\mathbf{s}} \in \mathbb{R}^n} \ \widetilde{J}(\widetilde{\mathbf{s}}) = \int_0^T \frac{1}{2} \widetilde{\mathbf{y}}(t)^T \widetilde{\mathbf{y}}(t) + \widetilde{\mathbf{c}}(t)^T \widetilde{\mathbf{y}}(t) dt + \frac{1}{2} \widetilde{\mathbf{y}}(T)^T \widetilde{\mathbf{y}}(T) + \widetilde{\mathbf{c}}_T^T \widetilde{\mathbf{y}}(T) + \frac{\beta}{2} \widetilde{\mathbf{s}}^T \widetilde{\mathbf{s}}, \tag{17a}$$

where for given $\widetilde{\mathbf{s}}$ the state $\widetilde{\mathbf{y}} \in H^1(0, T; \mathbb{R}^n)$ solves

$$\frac{d}{dt} \widetilde{\mathbf{y}}(t) + \mathbf{\Lambda} \widetilde{\mathbf{y}}(t) = \widetilde{\mathbf{f}}(t), \qquad\qquad t \in (0, T) \tag{17b}$$

$$\widetilde{\mathbf{y}}(0) = \widetilde{\mathbf{s}}. \tag{17c}$$

Because $\mathbf{s}^T \mathbf{M} \mathbf{s} = \widetilde{\mathbf{s}}^T \widetilde{\mathbf{s}}$, the $\mathbf{M}$-inner product in the $\mathbf{s}$ variables becomes the standard Euclidean inner product in the $\widetilde{\mathbf{s}}$ variables. The gradient of $\widetilde{J}$ in (17a) is given by

$$\nabla \widetilde{J}(\widetilde{\mathbf{s}}) = \widetilde{\mathbf{p}}(0) + \beta \widetilde{\mathbf{s}}, \tag{18}$$

where $\widetilde{\mathbf{p}}$ is the solution of the adjoint equation

$$-\frac{d}{dt} \widetilde{\mathbf{p}}(t) + \mathbf{\Lambda} \widetilde{\mathbf{p}}(t) = \widetilde{\mathbf{y}}(t) + \widetilde{\mathbf{c}}(t), \qquad\qquad t \in (0, T), \tag{19a}$$

$$\widetilde{\mathbf{p}}(T) = \widetilde{\mathbf{y}}(T) + \widetilde{\mathbf{c}}_T. \tag{19b}$$

Since $\mathbf{\Lambda}$ is diagonal, (17b, 17c) and (19) decouple into independent scalar problems, respectively. The ith component of the solution of the state equation (17b, 17c) is

$$\widetilde{\mathbf{y}}_i(t) = e^{-\lambda_i t} \widetilde{\mathbf{s}}_i + \int_0^t e^{-\lambda_i(t-\tau)} \widetilde{\mathbf{f}}_i(\tau) d\tau \tag{20}$$

and the ith component of the adjoint solution is given by

---

[1]Note $\mathbf{V}_n^T \mathbf{M} \mathbf{V}_n = \mathbf{I}$ implies that $\mathbf{V}_n$ is invertible and $\mathbf{V}_n \mathbf{V}_n^T \mathbf{M} = \mathbf{I}$. Thus, $\widetilde{\mathbf{y}}(t) = \mathbf{V}_n^T \mathbf{M} \mathbf{y}(t)$.

$$\widetilde{\mathbf{p}}_i(0) = e^{-\lambda_i T}[\widetilde{\mathbf{y}}_i(T) + \widetilde{\mathbf{c}}_{T,i}] + \int_0^T e^{-\lambda_i \tau}[\widetilde{\mathbf{y}}_i(\tau) + \widetilde{\mathbf{c}}_i(\tau)]d\tau$$

$$= h(\lambda_i)\widetilde{\mathbf{s}}_i + \widetilde{\mathbf{b}}_i, \tag{21}$$

where the function $h : \mathbb{R} \to \mathbb{R}$ is defined by

$$h(\lambda) = \begin{cases} e^{-2\lambda T} + \left(1 - e^{-2\lambda T}\right)\frac{1}{2\lambda}, & \text{if } \lambda \neq 0, \\ 1 + T, & \text{if } \lambda = 0, \end{cases} \tag{22}$$

and $\widetilde{\mathbf{b}}_i$ is the affine part, which depends on $\widetilde{\mathbf{c}}_i$, $\widetilde{\mathbf{c}}_{T,i}$, $\widetilde{\mathbf{f}}_i$, but is independent of $\widetilde{\mathbf{s}}_i$. Inserting (21) into (18) shows that the ith component of the gradient is

$$\frac{\partial}{\partial \widetilde{\mathbf{s}}_i} \widetilde{J}(\widetilde{\mathbf{s}}) = \left(h(\lambda_i) + \beta\right)\widetilde{\mathbf{s}}_i + \widetilde{\mathbf{b}}_i. \tag{23}$$

The Hessian of $\widetilde{J}$ is diagonal and independent of $\widetilde{\mathbf{s}}$,

$$\nabla^2 \widetilde{J} = \widetilde{\mathbf{H}} + \beta \mathbf{I},$$

where

$$\widetilde{\mathbf{H}} = \text{diag}(\widetilde{\mathbf{H}}_{11}, \ldots, \widetilde{\mathbf{H}}_{nn})$$

and

$$\widetilde{\mathbf{H}}_{ii} = h(\lambda_i). \tag{24}$$

The application of $\widetilde{\mathbf{H}}$ to a vector $\widetilde{\mathbf{s}}$ requires solutions of homogeneous versions of (17b, 17c) and (19). Specifically,

$$\widetilde{\mathbf{H}}\,\widetilde{\mathbf{s}} = \left(h(\lambda_i)\widetilde{\mathbf{s}}_i\right)_{i=1}^n = \widetilde{\mathbf{q}}(0), \tag{25}$$

where $\widetilde{\mathbf{q}}$ is obtained by first solving

$$\frac{d}{dt}\widetilde{\mathbf{z}}(t) + \mathbf{\Lambda}\widetilde{\mathbf{z}}(t) = 0, \qquad\qquad t \in (0, T), \tag{26a}$$

$$\widetilde{\mathbf{z}}(0) = \widetilde{\mathbf{s}}, \tag{26b}$$

and then

$$-\frac{d}{dt}\widetilde{\mathbf{q}}(t) + \mathbf{\Lambda}\widetilde{\mathbf{q}}(t) = \widetilde{\mathbf{z}}(t), \qquad\qquad t \in (0, T), \tag{27a}$$

$$\widetilde{\mathbf{q}}(T) = \widetilde{\mathbf{z}}(T). \tag{27b}$$

The following lemma from [8, L. 4.1] collects the properties of the function h. These will be used later in this paper.

Lemma 1 *For any $T > 0$, the function $h : \mathbb{R} \to \mathbb{R}$ defined by (22) is positive, convex, monotonically decreasing, and converges to zero as $\lambda \to \infty$.*

Sequential Hessian Approximation Using the Basic ROM

Let $\widetilde{\mathbf{s}}^{(0)}$ be given. The initial gradient is

$$\nabla \widetilde{J}(\widetilde{\mathbf{s}}^{(0)}) = (\widetilde{\mathbf{H}} + \beta \mathbf{I})\widetilde{\mathbf{s}}^{(0)} + \widetilde{\mathbf{b}}$$

according to (23) and (24). In step $\ell = 0$, the basic ROM for (26) and (27) is $\widetilde{\mathbf{V}} = [\mathbf{e}_1, \dots, \mathbf{e}_k] \in \mathbb{R}^{n \times k}$, which means that instead of solving all n equations (26) and all n equations (27), we only solve the first k equations in (26) and in (27), respectively. The approximate Hessian is

$$\widetilde{\mathbf{H}}_{\text{bsc}}^{(0)} + \beta \mathbf{I},$$

where

$$\widetilde{\mathbf{H}}_{\text{bsc}}^{(0)} = \text{diag}(\widetilde{\mathbf{H}}_{11}, \dots, \widetilde{\mathbf{H}}_{kk}, 0, \dots, 0). \tag{28}$$

The monotonicity of h, Lemma 1, and (24) imply that $\widetilde{\mathbf{H}}_{\text{bsc}}^{(0)}$ is the best rank-k approximation of $\widetilde{\mathbf{H}}$.

The solution of

$$\left(\widetilde{\mathbf{H}}_{\text{bsc}}^{(0)} + \beta \mathbf{I}\right) \Delta \widetilde{\mathbf{s}} = -\nabla \widetilde{J}(\widetilde{\mathbf{s}}^{(0)})$$

is

$$\Delta \widetilde{\mathbf{s}} = \begin{cases} -\left(\nabla \widetilde{J}(\widetilde{\mathbf{s}}^{(0)})\right)_i / (\widetilde{\mathbf{H}}_{ii} + \beta), & i = 1, \dots, k, \\ -\left(\nabla \widetilde{J}(\widetilde{\mathbf{s}}^{(0)})\right)_i / \beta, & i > k. \end{cases}$$

This leads to

$$\widetilde{\mathbf{s}}^{(1)} = \widetilde{\mathbf{s}}^{(0)} + \Delta \widetilde{\mathbf{s}} = \begin{cases} -\widetilde{\mathbf{b}}_i / (\widetilde{\mathbf{H}}_{ii} + \beta) = \widetilde{\mathbf{s}}_i^*, & i = 1, \dots, k, \\ -(\widetilde{\mathbf{H}}_{ii}\widetilde{\mathbf{s}}_i^{(0)} + \widetilde{\mathbf{b}}_i) / \beta, & i > k, \end{cases}$$

and

$$\nabla \widetilde{J}(\widetilde{\mathbf{s}}^{(1)}) = \begin{cases} 0, & i = 1, \dots, k, \\ -(\widetilde{\mathbf{H}}_{ii} / \beta)\left(\nabla \widetilde{J}(\widetilde{\mathbf{s}}^{(0)})\right)_i, & i > k. \end{cases}$$

After the initial step, the first k components of $\widetilde{\mathbf{s}}^{(1)}$ are exact, and the first k components of the gradient are zero. Thus, in step $\ell = 1$, we can repeat the same procedure on the subsystem with indices $k + 1$ to n. The ROM now is $\widetilde{\mathbf{V}} = [\mathbf{e}_{k+1}, \dots, \mathbf{e}_{2k}] \in \mathbb{R}^{n \times k}$, which results in the approximate Hessian $\widetilde{\mathbf{H}}_{\text{bsc}}^{(1)} + \beta \mathbf{I}$ with $\widetilde{\mathbf{H}}_{\text{bsc}}^{(1)} = \text{diag}(0, \dots, 0, \widetilde{\mathbf{H}}_{k+1,k+1}, \dots, \widetilde{\mathbf{H}}_{2k,2k}, 0, \dots, 0)$.

The solution of $\left(\widetilde{\mathbf{H}}_{bsc}^{(1)} + \beta \mathbf{I}\right) \Delta \widetilde{\mathbf{s}} = -\nabla \widetilde{J}(\widetilde{\mathbf{s}}^{(1)})$ is

$$\Delta \widetilde{\mathbf{s}} = \begin{cases} 0, & i = 1, \ldots, k, \\ -\left(\nabla \widetilde{J}(\widetilde{\mathbf{s}}^{(1)})\right)_i / (\widetilde{\mathbf{H}}_{ii} + \beta), & i = k+1, \ldots, 2k, \\ -\left(\nabla \widetilde{J}(\widetilde{\mathbf{s}}^{(1)})\right)_i / \beta, & i > 2k. \end{cases}$$

This gives

$$\widetilde{\mathbf{s}}^{(2)} = \widetilde{\mathbf{s}}^{(1)} + \Delta \widetilde{\mathbf{s}} = \begin{cases} -\widetilde{\mathbf{b}}_i / (\widetilde{\mathbf{H}}_{ii} + \beta) = \widetilde{\mathbf{s}}_i^*, & i = 1, \ldots, 2k, \\ -(\widetilde{\mathbf{H}}_{ii} \widetilde{\mathbf{s}}_i^{(1)} + \widetilde{\mathbf{b}}_i) / \beta, & i > 2k \end{cases}$$

and

$$\nabla \widetilde{J}(\widetilde{\mathbf{s}}^{(2)}) = \begin{cases} 0, & i = 1, \ldots, 2k, \\ -(\widetilde{\mathbf{H}}_{ii} / \beta) \left(\nabla \widetilde{J}(\widetilde{\mathbf{s}}^{(1)})\right)_i, & i > 2k. \end{cases}$$

More generally, we have the following result.

Theorem 1 *If the ROM in the ℓth step is $\widetilde{\mathbf{V}} = [\mathbf{e}_{\ell k+1}, \ldots, \mathbf{e}_{(\ell+1)k}] \in \mathbb{R}^{n \times k}$ with resulting approximate Hessian $\widetilde{\mathbf{H}}_{bsc}^{(\ell)} + \beta \mathbf{I}$, where*

$$\widetilde{\mathbf{H}}_{bsc}^{(\ell)} = diag(0, \ldots, 0, \widetilde{\mathbf{H}}_{\ell k+1, \ell k+1}, \ldots, \widetilde{\mathbf{H}}_{(\ell+1)k, (\ell+1)k}, 0, \ldots, 0), \qquad (29)$$

and $\Delta \widetilde{\mathbf{s}}$ solves $\left(\widetilde{\mathbf{H}}_{bsc}^{(\ell)} + \beta \mathbf{I}\right) \Delta \widetilde{\mathbf{s}} = -\nabla \widetilde{J}(\widetilde{\mathbf{s}}^{(\ell)})$, then the new iterate $\widetilde{\mathbf{s}}^{(\ell+1)} = \widetilde{\mathbf{s}}^{(\ell)} + \Delta \widetilde{\mathbf{s}}$ is

$$\widetilde{\mathbf{s}}^{(\ell+1)} = \begin{cases} -\widetilde{\mathbf{b}}_i / (\widetilde{\mathbf{H}}_{ii} + \beta) = \widetilde{\mathbf{s}}_i^*, & i = 1, \ldots, (\ell+1)k, \\ -(\widetilde{\mathbf{H}}_{ii} \widetilde{\mathbf{s}}_i^{(\ell)} + \widetilde{\mathbf{b}}_i) / \beta, & i > (\ell+1)k \end{cases}$$

and

$$\nabla \widetilde{J}(\widetilde{\mathbf{s}}^{(\ell+1)}) = \begin{cases} 0, & i = 1, \ldots, (\ell+1)k, \\ -(\widetilde{\mathbf{H}}_{ii} / \beta) \left(\nabla \widetilde{J}(\widetilde{\mathbf{s}}^{(\ell)})\right)_i, & i > (\ell+1)k. \end{cases}$$

The proof of Theorem 1 is by straightforward induction and is omitted because of page limitations.

Theorem 1 shows that

$$\left(\nabla \widetilde{J}(\widetilde{\mathbf{s}}^{(\ell)})\right)_i = (-\widetilde{\mathbf{H}}_{ii} / \beta)^\ell \left(\nabla \widetilde{J}(\widetilde{\mathbf{s}}^{(0)})\right)_i, \quad i > \ell k,$$

which implies

$$\left| \left(\nabla \widetilde{J}(\widetilde{\mathbf{s}}^{(\ell)})\right)_i \right| = \left(\frac{\widetilde{\mathbf{H}}_{ii}}{\beta}\right)^\ell \left| \left(\nabla \widetilde{J}(\widetilde{\mathbf{s}}^{(0)})\right)_i \right| \gg \left| \left(\nabla \widetilde{J}(\widetilde{\mathbf{s}}^{(0)})\right)_i \right|, \quad \text{if } \widetilde{\mathbf{H}}_{ii} > \beta, \quad i > \ell k. \tag{30}$$

To remedy this potential increase in gradient components, we only solve the subsystem

$$\left(\widetilde{\mathbf{H}}_{bsc}^{(\ell)} + \beta \mathbf{I}\right)_i \Delta \widetilde{\mathbf{s}}_i = -\left(\nabla \widetilde{J}(\widetilde{\mathbf{s}}^{(\ell)})\right)_i, \quad i = \ell k+1, \ldots, (\ell+1)k \tag{31}$$

and set $\Delta\widetilde{\mathbf{s}}_i = 0$ for $i \leqslant \ell k$ and $i > (\ell+1)k$.

Theorem 2 *If the ROM in the ℓth step is* $\widetilde{\mathbf{V}} = [\mathbf{e}_{\ell k+1}, \ldots, \mathbf{e}_{(\ell+1)k}] \in \mathbb{R}^{n \times k}$ *with resulting approximate Hessian* $\widetilde{\mathbf{H}}_{bsc}^{(\ell)} + \beta\mathbf{I}$, *where* $\widetilde{\mathbf{H}}_{bsc}^{(\ell)}$ *is given by* (29), *and* $\Delta\widetilde{\mathbf{s}}$ *solves* (31), *then the resulting iterate* $\widetilde{\mathbf{s}}^{(\ell+1)} = \widetilde{\mathbf{s}}^{(\ell)} + \Delta\widetilde{\mathbf{s}}$ *is*

$$\widetilde{\mathbf{s}}^{(\ell+1)} = \begin{cases} -\widetilde{\mathbf{b}}_i / (\widetilde{\mathbf{H}}_{ii} + \beta) = \widetilde{\mathbf{s}}_i^*, & i = 1, \ldots, (\ell+1)k, \\ \widetilde{\mathbf{s}}_i^{(0)}, & i > (\ell+1)k \end{cases}$$

and

$$\nabla\widetilde{J}(\widetilde{\mathbf{s}}^{(\ell+1)}) = \begin{cases} 0, & i = 1, \ldots, (\ell+1)k, \\ \left(\nabla\widetilde{J}(\widetilde{\mathbf{s}}^{(0)})\right)_i, & i > (\ell+1)k. \end{cases}$$

The proof of Theorem 2 is by straightforward induction and is omitted because of page limitations.

By design, the norms of the gradients corresponding to the sequence generated by (31) are monotonically decreasing,

$$\|\nabla\widetilde{J}(\widetilde{\mathbf{s}}^{(\ell+1)})\| \leqslant \|\nabla\widetilde{J}(\widetilde{\mathbf{s}}^{(\ell)})\|, \tag{32}$$

with '$<$' if the gradient at the initial iterate has a nonzero component $\left(\nabla\widetilde{J}(\widetilde{\mathbf{s}}^{(0)})\right)_i$, $i \in \{\ell k+1, \ldots, (\ell+1)k\}$.

Sequential Hessian Approximation Using the Augmented ROM

The basic ROM described in the previous subsection generates the best rank k approximation of the Hessian, but neglects the right-hand side $\nabla\widetilde{J}(\widetilde{\mathbf{s}}^{(\ell)})$. To improve the quality of the basic ROM, we proposed in [8] to augment it by the right-hand side vector. We describe the ROM augmentation in the sequential application for step $\ell \in \mathbb{N}_0$.

Let $\widetilde{\mathbf{s}}^{(\ell)}$ be given such that

$$\left(\nabla\widetilde{J}(\widetilde{\mathbf{s}}^{(\ell)})\right)_i = 0, \quad i = 1, \ldots, \ell k. \tag{33}$$

The basic ROM $\widetilde{\mathbf{V}} = [\mathbf{e}_{\ell k+1}, \ldots, \mathbf{e}_{(\ell+1)k}]$ is augmented by

$$\widetilde{\mathbf{g}} = \nabla\widetilde{J}(\widetilde{\mathbf{s}}^{(\ell)}),$$

(note that $\widetilde{\mathbf{g}} = \widetilde{\mathbf{g}}^{(\ell)}$, but we drop the superscript to simplify notation) and is given by

$$\widetilde{\mathbf{V}}^{\mathrm{aug}} = [\mathbf{e}_{\ell k+1}, \ldots, \mathbf{e}_{(\ell+1)k}, \widetilde{\mathbf{v}}] \in \mathbb{R}^{n \times (k+1)}, \tag{34a}$$

where

$$\widetilde{\mathbf{v}} = \left(0, \ldots, 0, \widetilde{\mathbf{g}}_{(\ell+1)k+1}, \ldots, \widetilde{\mathbf{g}}_n\right)^T \Big/ \left(\sum_{i=(\ell+1)k+1}^{n} \widetilde{\mathbf{g}}_i^2\right)^{1/2}. \tag{34b}$$

Now we project the systems (26) and (27) with $\widetilde{\mathbf{V}}^{\mathrm{aug}}$. Assume $\sum_{i=(\ell+1)k+1}^{n} \widetilde{\mathbf{g}}_i^2 > 0$, i.e., $\widetilde{\mathbf{g}} \notin \widetilde{\mathbf{V}} = [\mathbf{e}_{\ell k+1}, \ldots, \mathbf{e}_{(\ell+1)k}]$, and define

$$\lambda(\widetilde{\mathbf{g}}) = \widetilde{\mathbf{v}}^T \Lambda \widetilde{\mathbf{v}} = \sum_{i=(\ell+1)k+1}^{n} \lambda_i \widetilde{\mathbf{g}}_i^2 \Big/ \sum_{i=(\ell+1)k+1}^{n} \widetilde{\mathbf{g}}_i^2. \tag{35}$$

An extension of [8, Eq. (4.26)] shows that the Hessian approximation resulting from the subspace specified in (34) is given by

$$\begin{aligned}
\widetilde{\mathbf{H}}^{(\ell)}_{\mathrm{aug}}\,\widetilde{\mathbf{s}} &= \sum_{i=\ell k+1}^{(\ell+1)k} \widetilde{\mathbf{H}}_{ii}\,\widetilde{\mathbf{s}}_i\,\mathbf{e}_i \\
&\quad + h\big(\lambda(\widetilde{\mathbf{g}})\big) \frac{\sum_{i=(\ell+1)k+1}^{n} \widetilde{\mathbf{g}}_i \widetilde{\mathbf{s}}_i}{\sum_{i=(\ell+1)k+1}^{n} \widetilde{\mathbf{g}}_i^2} \left(0, \ldots, 0, \widetilde{\mathbf{g}}_{(\ell+1)k+1}, \ldots, \widetilde{\mathbf{g}}_n\right)^T
\end{aligned} \tag{36}$$

for all $\widetilde{\mathbf{s}} \in \mathbb{R}^n$, where h is the function defined in (22). Since (36) holds for all $\widetilde{\mathbf{s}} \in \mathbb{R}^n$,

$$\widetilde{\mathbf{H}}^{(\ell)}_{\mathrm{aug}} = \begin{pmatrix} 0 & & & & & & & \\ & \ddots & & & & & & \\ & & 0 & & & & & \\ & & & \widetilde{\mathbf{H}}_{\ell k+1, \ell k+1} & & & & \\ & & & & \ddots & & & \\ & & & & & \widetilde{\mathbf{H}}_{(\ell+1)k,(\ell+1)k} & & \\ & & & & & & 0 & \\ & & & & & & & \ddots \\ & & & & & & & & 0 \end{pmatrix}$$

$$+ \frac{h\big(\lambda(\widetilde{\mathbf{g}})\big)}{\sum_{i=(\ell+1)k+1}^{n} \widetilde{\mathbf{g}}_i^2} \begin{pmatrix} 0 \\ \vdots \\ 0 \\ \widetilde{\mathbf{g}}_{(\ell+1)k+1} \\ \vdots \\ \widetilde{\mathbf{g}}_n \end{pmatrix} \left(0, \ldots, 0, \widetilde{\mathbf{g}}_{(\ell+1)k+1}, \ldots, \widetilde{\mathbf{g}}_n\right). \tag{37}$$

Given the Hessian approximation, (37) we solve

$$(\widetilde{\mathbf{H}}^{(\ell)}_{\mathrm{aug}} + \beta \mathbf{I})\Delta \widetilde{\mathbf{s}} = -\nabla \widetilde{J}(\widetilde{\mathbf{s}}^{(\ell)}) \tag{38}$$

and set $\widetilde{\mathbf{s}}^{(\ell+1)} = \widetilde{\mathbf{s}}^{(\ell)} + \Delta\widetilde{\mathbf{s}}$.

Theorem 3 *Let* $\widetilde{\mathbf{g}} = \nabla\widetilde{J}(\widetilde{\mathbf{s}}^{(\ell)})$ *with* $\widetilde{\mathbf{g}}_i = 0$, $i = 1, \ldots, \ell k$, *and* $\sum_{j=(\ell+1)k+1}^n \widetilde{\mathbf{g}}_j^2 > 0$, *and let* $\widetilde{\mathbf{H}}_{aug}^{(\ell)}$ *be given by* (37). *The solution of* (38) *is given by*

$$
\Delta\widetilde{\mathbf{s}}_i = \begin{cases} 0, & i = 1, \ldots, \ell k, \\ -\left(\widetilde{\mathbf{H}}_{ii} + \beta\right)^{-1}\left(\nabla\widetilde{J}(\widetilde{\mathbf{s}}^{(\ell)})\right)_i, & i = \ell k + 1, \ldots, (\ell+1)k, \\ -\left(h\big(\lambda(\widetilde{\mathbf{g}})\big) + \beta\right)^{-1}\left(\nabla\widetilde{J}(\widetilde{\mathbf{s}}^{(\ell)})\right)_i, & i = (\ell+1)k + 1, \ldots, n, \end{cases} \tag{39}
$$

the resulting iterate $\widetilde{\mathbf{s}}^{(\ell+1)} = \widetilde{\mathbf{s}}^{(\ell)} + \Delta\widetilde{\mathbf{s}}$ *is*

$$
\widetilde{\mathbf{s}}^{(\ell+1)} = \begin{cases} -\widetilde{\mathbf{b}}_i/(\widetilde{\mathbf{H}}_{ii} + \beta) = \widetilde{\mathbf{s}}_i^*, & i = 1, \ldots, (\ell+1)k, \\ \dfrac{h(\lambda(\widetilde{\mathbf{g}})) - \widetilde{\mathbf{H}}_{ii}}{h(\lambda(\widetilde{\mathbf{g}})) + \beta}\widetilde{\mathbf{s}}_i^{(\ell)} - \dfrac{\widetilde{\mathbf{b}}_i}{h(\lambda(\widetilde{\mathbf{g}})) + \beta}, & i > (\ell+1)k, \end{cases} \tag{40}
$$

and

$$
\nabla\widetilde{J}(\widetilde{\mathbf{s}}^{(\ell+1)}) = \begin{cases} 0, & i = 1, \ldots, (\ell+1)k, \\ \dfrac{h(\lambda(\widetilde{\mathbf{g}})) - \widetilde{\mathbf{H}}_{ii}}{h(\lambda(\widetilde{\mathbf{g}})) + \beta}\left(\nabla\widetilde{J}(\widetilde{\mathbf{s}}^{(\ell)})\right)_i, & i > (\ell+1)k. \end{cases} \tag{41}
$$

Proof The identity (39) can be proven as in [8, Lemma 4.2]. Then the identity (40) is obtained by inserting (39) into $\widetilde{\mathbf{s}}^{(\ell+1)} = \widetilde{\mathbf{s}}^{(\ell)} + \Delta\widetilde{\mathbf{s}}$, and (41) is obtained by inserting (40) into (23). □

If $\|\nabla\widetilde{J}(\widetilde{\mathbf{s}}^{(\ell+1)})\|$ is not small enough, we repeat the procedure with ℓ replaced by $\ell + 1$. Note that the first $(\ell+1)k$ components of $\nabla\widetilde{J}(\widetilde{\mathbf{s}}^{(\ell+1)})$ are zero and the assumption (33) is satisfied if ℓ is replaced by $\ell + 1$.

Theorem 4 *Under the assumptions of Theorem 3,*

$$
\|\nabla\widetilde{J}(\widetilde{\mathbf{s}}^{(\ell+1)})\| \leqslant \eta^{(\ell)}\|\nabla\widetilde{J}(\widetilde{\mathbf{s}}^{(\ell)})\|,
$$

where

$$
\eta^{(\ell)} = \frac{\big(h(\lambda_{(\ell+1)k+1})\big)^2 - \big(h(\lambda(\widetilde{\mathbf{g}}))\big)^2}{\big(h(\lambda(\widetilde{\mathbf{g}})) + \beta\big)^2} \sum_{i=(\ell+1)k+1}^n \left(\nabla\widetilde{J}(\widetilde{\mathbf{s}}^{(\ell)})\right)_i^2 \Big/ \sum_{i=\ell k+1}^n \left(\nabla\widetilde{J}(\widetilde{\mathbf{s}}^{(\ell)})\right)_i^2.
$$

Before we prove this result, note that $\eta^{(\ell)} < 1$ if

$$
\frac{\big(h(\lambda_{(\ell+1)k+1})\big)^2 - \big(h(\lambda(\widetilde{\mathbf{g}}))\big)^2}{\big(h(\lambda(\widetilde{\mathbf{g}})) + \beta\big)^2} < \left(1 + \sum_{i=\ell k+1}^{(\ell+1)k} \left(\nabla\widetilde{J}(\widetilde{\mathbf{s}}^{(\ell)})\right)_i^2 \Big/ \sum_{i=(\ell+1)k+1}^n \left(\nabla\widetilde{J}(\widetilde{\mathbf{s}}^{(\ell)})\right)_i^2\right),
$$

i.e., if the relative change in h^2 on $[\lambda_{(\ell+1)k+1}, \lambda_n]$ is small compared to the relative size of the gradient components $\left(\nabla\widetilde{J}(\mathbf{s}^{(\ell)})\right)_i^2, i = \ell k + 1, \dots, (\ell+1)k$, that are made zero in step ℓ. This is the case in our numerical example shown in Sect. 6.

Proof Define $\widetilde{\mathbf{g}} = \nabla\widetilde{J}(\mathbf{s}^{(\ell)})$ and recall that $\widetilde{\mathbf{g}}_i = 0, i = 1, \dots, \ell k$.

The identities (41) and (24) lead to

$$
\begin{aligned}
\|\nabla\widetilde{J}(\mathbf{s}^{(\ell+1)})\|^2 &= \sum_{i=(\ell+1)k+1}^{n} \left(\frac{h(\lambda(\widetilde{\mathbf{g}})) - h(\lambda_i)}{h(\lambda(\widetilde{\mathbf{g}})) + \beta}\right)^2 \widetilde{\mathbf{g}}_i^2 \\
&= \frac{\left(h(\lambda(\widetilde{\mathbf{g}}))\right)^2}{\left(h(\lambda(\widetilde{\mathbf{g}})) + \beta\right)^2} \sum_{i=(\ell+1)k+1}^{n} \widetilde{\mathbf{g}}_i^2 - 2\frac{h(\lambda(\widetilde{\mathbf{g}}))}{\left(h(\lambda(\widetilde{\mathbf{g}})) + \beta\right)^2} \sum_{(\ell+1)k+1}^{n} h(\lambda_i)\,\widetilde{\mathbf{g}}_i^2 \\
&\quad + \frac{1}{\left(h(\lambda(\widetilde{\mathbf{g}})) + \beta\right)^2} \sum_{i=(\ell+1)k+1}^{n} \left(h(\lambda_i)\right)^2 \widetilde{\mathbf{g}}_i^2. \quad (42)
\end{aligned}
$$

Since $\lambda(\widetilde{\mathbf{g}})$ is a weighted arithmetic mean of $\lambda_{(\ell+1)k+1}, \dots, \lambda_n > 0$, it holds that $\lambda(\widetilde{\mathbf{g}}) \in [\lambda_{(\ell+1)k+1}, \lambda_n]$. Moreover, due to the convexity of h (cf. Lemma 1) Jensen's inequality,

$$
\sum_{i=(\ell+1)k+1}^{n} h(\lambda_i)\,\widetilde{\mathbf{g}}_i^2 \Big/ \sum_{j=(\ell+1)k+1}^{n} \widetilde{\mathbf{g}}_j^2 \geqslant h(\lambda(\widetilde{\mathbf{g}})), \quad (43)
$$

holds. Using (43) in the second sum in (42) and $0 < h(\lambda_i) \leqslant h(\lambda_{(\ell+1)k+1})$ for $i \geqslant (\ell+1)k + 1$ (by monotonicity of h, Lemma 1) in the third sum in (42) gives

$$
\begin{aligned}
\|\nabla\widetilde{J}(\mathbf{s}^{(\ell+1)})\|^2 &\leqslant \frac{\left(h(\lambda_{(\ell+1)k+1})\right)^2 - \left(h(\lambda(\widetilde{\mathbf{g}}))\right)^2}{\left(h(\lambda(\widetilde{\mathbf{g}})) + \beta\right)^2} \sum_{i=(\ell+1)k+1}^{n} \widetilde{\mathbf{g}}_i^2 \\
&\leqslant \eta^{(\ell)} \sum_{i=\ell k+1}^{n} \widetilde{\mathbf{g}}_i^2 = \eta^{(\ell)}\|\nabla\widetilde{J}(\mathbf{s}^{(\ell)})\|^2.
\end{aligned}
$$

$\square$

5 Algorithms Based on Sequential Reduced-Order Models

Subsections "Sequential Hessian Approximation Using the Basic ROM" and "Sequential Hessian Approximation Using the Augmented ROM" of Sect. 4 presented the basic idea of our sequential approach for the diagonalized case (17). In practice, one does not perform the variable transformations that lead to (17), but instead works with (1) directly. This section describes our sequential approach for (1) under the assumptions (3).

Throughout this section, we assume that assumptions (3) hold, and we let $\mathbf{V}_n \in \mathbb{R}^{n \times n}$ be the matrix of eigenvectors of $(\mathbf{A}, \mathbf{M})$ satisfying (15).

As outlined in subsection "Iteration" of Sect. 3, we solve a sequence of problems

$$(\widehat{\mathbf{H}}^{(\ell)} + \beta \mathbf{I}) \, \Delta \mathbf{s} = -\nabla J(\mathbf{s}^{(\ell)}), \tag{44}$$

and set $\mathbf{s}^{(\ell+1)} = \mathbf{s}^{(\ell)} + \Delta \mathbf{s}$. The ROM is either

$$\mathbf{V} = [(\mathbf{V}_n)_{\ell k+1}, \, \cdots, \, (\mathbf{V}_n)_{(\ell+1)k}] \in \mathbb{R}^{n \times k},$$

where $(\mathbf{V}_n)_i$ is the ith column of $\mathbf{V}_n$ or it is the augmentation of this basic ROM,

$$\mathbf{V} = [(\mathbf{V}_n)_{\ell k+1}, \, \cdots, \, (\mathbf{V}_n)_{(\ell+1)k}, \mathbf{v}] \in \mathbb{R}^{n \times (k+1)},$$

such that $\nabla J(\mathbf{s}^{(\ell)}) \in \text{range}(\mathbf{V})$. Once $\mathbf{V}$ is constructed, the reduced-order Hessian $\widehat{\mathbf{H}}^{(\ell)}$ is given by (14).

In subsections "Sequential Hessian Approximation Using the Basic ROM" and "Sequential Hessian Approximation Using the Augmented ROM" of Sect. 4 we have seen that if we solve the systems (44) exactly, then the first ℓk components of $\mathbf{V}_n^T \mathbf{M} \nabla J(\mathbf{s}^{(\ell)})$ are zero. Thus, in this case after a change of variables, solving the $n \times n$ system (44) is equivalent to solving an $(n - \ell k) \times (n - \ell k)$ subsystem. If the systems (44) are not solved exactly, this is not true and we use a projection to essentially remove the first ℓk equations and unknowns from (44). More precisely, let

$$\mathbf{V}^{\text{old}} = [(\mathbf{V}_n)_1, \, \cdots, \, (\mathbf{V}_n)_{\ell k}] \in \mathbb{R}^{n \times (\ell k)}$$

and define the projection

$$\mathbf{P} = \mathbf{I} - \mathbf{V}^{\text{old}}(\mathbf{V}^{\text{old}})^T \mathbf{M}. \tag{45}$$

Using (15), it is easy to verify that

$$\mathbf{P}^2 = \mathbf{P},$$
$$\mathbf{P}^* = \mathbf{M}^{-1}\mathbf{P}^T \mathbf{M} = \mathbf{P},$$

i.e., $\mathbf{P}$ is in fact a projection and orthogonal with respect to the inner product (2) (recall that $\mathbf{G} = \mathbf{M}$).

Instead of (44), we solve

$$\mathbf{P}^*(\widehat{\mathbf{H}}^{(\ell)} + \beta \mathbf{I})\mathbf{P}\Delta\widehat{\mathbf{s}} = -\mathbf{P}^*\nabla J(\mathbf{s}^{(\ell)}). \tag{46}$$

The next lemma shows that the system matrix in (46) is self-adjoint and positive definite on range($\mathbf{P}$). Therefore, the system (46) can be solved using the conjugate gradient (CG) method. Furthermore, (14) reveals that $\widehat{\mathbf{H}}^{(\ell)}$ has rank k if the basic ROM is used or $k + 1$ if the augmented ROM is used, which implies that the CG method applied to (46) converges in at most k or $k + 1$ iterations, see [8, Sect. 6.2] for details.

Lemma 2 *The matrix* $\mathbf{P}^*(\widehat{\mathbf{H}}^{(\ell)} + \beta\mathbf{I})\mathbf{P}$ *is self-adjoint with respect to the inner product* (2) *and positive definite on* range($\mathbf{P}$).

Proof Recall that $\mathbf{M} = \mathbf{G}$ and that the adjoint in the inner product (2) is $\mathbf{N}^* = \mathbf{M}^{-1}\mathbf{N}^T\mathbf{M}$. Using the self-adjointness of $\mathbf{P}$ and of $\widehat{\mathbf{H}}^{(\ell)}$ (see end of Sect. "Reduced-Order Model Hessian Approximation" or [8]), the self-adjointness of $\mathbf{P}^*(\widehat{\mathbf{H}}^{(\ell)} + \beta\mathbf{I})\mathbf{P}$ follows from

$$
\begin{aligned}
\left(\mathbf{P}^*(\widehat{\mathbf{H}}^{(\ell)} + \beta\mathbf{I})\mathbf{P}\right)^* &= \mathbf{M}^{-1}[\mathbf{P}^*(\widehat{\mathbf{H}}^{(\ell)} + \beta\mathbf{I})\mathbf{P}]^T\mathbf{M} = \mathbf{M}^{-1}\mathbf{P}^T(\widehat{\mathbf{H}}^{(\ell)} + \beta\mathbf{I})^T(\mathbf{P}^*)^T\mathbf{M} \\
&= \mathbf{M}^{-1}\mathbf{P}^T\mathbf{M}\mathbf{M}^{-1}(\widehat{\mathbf{H}}^{(\ell)} + \beta\mathbf{I})^T\mathbf{M}\mathbf{P}\mathbf{M}^{-1}\mathbf{M} \\
&= \mathbf{P}^*\mathbf{M}^{-1}(\widehat{\mathbf{H}}^{(\ell)} + \beta\mathbf{I})^T\mathbf{M}\mathbf{P} = \mathbf{P}^*(\widehat{\mathbf{H}}^{(\ell)} + \beta\mathbf{I})^*\mathbf{P} \\
&= \mathbf{P}^*(\widehat{\mathbf{H}}^{(\ell)} + \beta\mathbf{I})\mathbf{P}.
\end{aligned}
$$

Any $\mathbf{w} \in$ range($\mathbf{P}$) can be written as $\mathbf{w} = \mathbf{P}\mathbf{w}$. Thus,

$$
\langle \mathbf{w}, \mathbf{P}^*(\widehat{\mathbf{H}}^{(\ell)} + \beta\mathbf{I})\mathbf{P}\mathbf{w}\rangle = \langle \mathbf{P}\mathbf{w}, (\widehat{\mathbf{H}}^{(\ell)} + \beta\mathbf{I})\mathbf{P}\mathbf{w}\rangle = \langle \mathbf{w}, (\widehat{\mathbf{H}}^{(\ell)} + \beta\mathbf{I})\mathbf{w}\rangle \geq \beta\langle \mathbf{w}, \mathbf{w}\rangle
$$

for all $\mathbf{w} \in$ range($\mathbf{P}$), since $\widehat{\mathbf{H}}^{(\ell)}$ is positive semidefinite (see end of Sect. "Reduced-Order Model Hessian Approximation" or [8]) and $\beta > 0$. □

The first version of our sequential approach is summarized in Algorithm 5. This algorithm corresponds to the scenario in Theorem 1 if the basic ROM is used or in Theorem 3 if the augmented ROM is used.

Algorithm 5 (Optimization with Sequential ROM Hessians)

0. Given tol $\in (0, 1)$, $\mathbf{s}^{(0)}$, k.
1. Set $\mathbf{V}^{\text{old}} = []$, $\mathbf{P} = \mathbf{I}$.
2. Compute the matrix $\mathbf{V}_n$ of eigenvectors of $(\mathbf{A}, \mathbf{M})$ satisfying (15).
3. For $\ell = 0, 1, 2, \ldots, \ell_{\max}$

    a. If $\|\nabla J(\mathbf{s}^{(\ell)})\|_{\mathbf{G}} < $ tol, return $\mathbf{s}^{(\ell)}$.

    b. Choose ROM $\mathbf{V}^{(\ell)}$:

       i. Either as the basic ROM with columns $\ell k + 1$ to $(\ell + 1)k$ from $\mathbf{V}_n$ which gives

$$
\mathbf{V}^{(\ell)} = [(\mathbf{V}_n)_{\ell k+1}, \cdots, (\mathbf{V}_n)_{(\ell+1)k}] \in \mathbb{R}^{n \times k},
$$

       where $(\mathbf{V}_n)_i$ is the ith column of $\mathbf{V}_n$,

      ii. or as this basic ROM augmented by $\mathbf{P}^*\nabla J(\mathbf{s}^{(\ell)})$ which gives

$$
\mathbf{V}^{(\ell)} = [(\mathbf{V}_n)_{\ell k+1}, \cdots, (\mathbf{V}_n)_{(\ell+1)k}, \mathbf{v}] \in \mathbb{R}^{n \times (k+1)},
$$

       where $\mathbf{v}$ is defined by (16b) with $\mathbf{g} = \mathbf{P}^*\nabla J(\mathbf{s}^{(\ell)})$.

c. For the Hessian approximation $\widehat{\mathbf{H}}^{(\ell)}$ obtained by $\mathbf{V}^{(\ell)}$, solve

$$\mathbf{P}^*(\widehat{\mathbf{H}}^{(\ell)} + \beta\mathbf{I})\mathbf{P}\varDelta\widehat{\mathbf{s}} = -\mathbf{P}^*\nabla J(\mathbf{s}^{(\ell)}). \tag{47}$$

d. Update $\mathbf{s}^{(\ell+1)} = \mathbf{s}^{(\ell)} + \mathbf{P}\varDelta\widehat{\mathbf{s}}$.
e. Set $\mathbf{V}^{\text{old}} = [\mathbf{V}^{\text{old}}, \mathbf{V}_1^{(\ell)}, \cdots, \mathbf{V}_k^{(\ell)}] \in \mathbb{R}^{n \times (\ell+1)k}$ where $\mathbf{V}_i^{(\ell)}$ is the ith column of $\mathbf{V}^{(\ell)}$.
f. Let $\mathbf{P} = \mathbf{I} - \mathbf{V}^{\text{old}}(\mathbf{V}^{\text{old}})^T\mathbf{M}$ be the $\mathbf{M}$-orthogonal projection on the complement of $\mathbf{V}^{\text{old}}$.

Remark 1   1. If the systems (47) are solved exactly, then Algorithm 5 terminates after at most $n/k - 1$ iterations with the exact solution.
2. As mentioned before, (47) can be solved using the conjugate gradient (CG) method. We choose

$$\text{tol} \cdot \frac{k}{n} \cdot c_{\text{safe}} \tag{48}$$

as the residual termination tolerance for the CG method where $c_{\text{safe}} \in (0, 1)$ is a safety factor and tol the tolerance of Step 3a in Algorithm 5. With this tolerance Algorithm 5 terminates after at most $n/k - 1$ iterations.
3. In practice, one should not compute the entire matrix $\mathbf{V}_n$ of eigenvectors of $(\mathbf{A}, \mathbf{M})$, but only k eigenvectors at a time, using an iterative eigensolver such as ARPACK [11] or `eigs` in Matlab.
4. Similarly, $\mathbf{P}$ is never computed explicitly. Solution of (47) by the CG algorithm only required application of $\mathbf{P}$ to vectors.

As we have seen in (30), the gradient norm can increase if Algorithm 5 is used with the basic ROM. In subsection "Sequential Hessian Approximation Using the Basic ROM", of Sect. 4 we proposed to compute the step by solving (31). In the context of the original problem, this means we apply another projection

$$\mathbf{Q} = \mathbf{V}^{(\ell)}(\mathbf{V}^{(\ell)})^T\mathbf{M}$$

and solve

$$\mathbf{Q}^*(\widehat{\mathbf{H}}^{(\ell)} + \beta\mathbf{I})\mathbf{Q}\varDelta\mathbf{s} = -\mathbf{Q}^*\nabla J(\mathbf{s}^{(\ell)}). \tag{49}$$

Using (15), it is easy to verify that

$$\mathbf{Q}^2 = \mathbf{Q},$$
$$\mathbf{Q}^* = \mathbf{M}^{-1}\mathbf{Q}^T\mathbf{M} = \mathbf{Q},$$

i.e., that $\mathbf{Q}$ is an orthogonal projection, and using

$$(\mathbf{V}^{\text{old}})^T\mathbf{M}\mathbf{V}^{(\ell)} = \mathbf{0},$$

it is easy to verify that

$$\mathbf{PQ} = \mathbf{QP} = \mathbf{Q} \quad \text{and} \quad \mathbf{P}^*\mathbf{Q}^* = \mathbf{Q}^*\mathbf{P}^* = \mathbf{Q}^*.$$

Thus, (49) is equivalent to

$$\mathbf{P}^*\mathbf{Q}^*(\widehat{\mathbf{H}}^{(\ell)} + \beta\mathbf{I})\mathbf{QP}\Delta\mathbf{s} = -\mathbf{P}^*\mathbf{Q}^*\nabla J(\mathbf{s}^{(\ell)})$$

Applying the projection $\mathbf{Q}$ corresponds to, after a variable transformation, setting all components except the components $\ell k + 1, \ldots, (\ell + 1)k$ to zero (for augmented ROM: except the components $\ell k + 1, \ldots, (\ell + 1)k + 1$).

Lemma 3 *The matrix* $\mathbf{Q}^*(\widehat{\mathbf{H}}^{(\ell)} + \beta\mathbf{I})\mathbf{Q}$ *is self-adjoint with respect to the inner product* (2) *and positive definite on* range($\mathbf{Q}$).

The proof is nearly identical to that of Lemma 2.

These modifications lead to Algorithm 6 below. This algorithm corresponds to the scenario in Theorem 2 if the basic ROM is used, but can also be applied to the augmented ROM.

Algorithm 6 (Optimization with Sequential ROM Hessians and Projection)

0. Given tol $\in (0, 1)$, $\mathbf{s}^{(0)}$, k.
1. Set $\mathbf{V}^{\text{old}} = []$, $\mathbf{P} = \mathbf{I}$, $\mathbf{Q} = \mathbf{I}$.
2. Compute the matrix $\mathbf{V}_n$ of eigenvectors of $(\mathbf{A}, \mathbf{M})$ satisfying (15).
3. For $\ell = 0, 1, 2, \ldots, \ell_{\max}$

    a. If $\|\nabla J(\mathbf{s}^{(\ell)})\|_{\mathbf{G}} < \text{tol}$, return $\mathbf{s}^{(\ell)}$.
    b. Choose ROM $\mathbf{V}^{(\ell)}$:
        i. Either as the basic ROM with columns $\ell k + 1$ to $(\ell + 1)k$ from $\mathbf{V}_n$ which gives

$$\mathbf{V}^{(\ell)} = [(\mathbf{V}_n)_{\ell k+1}, \cdots, (\mathbf{V}_n)_{(\ell+1)k}] \in \mathbb{R}^{n \times k}$$

        where $(\mathbf{V}_n)_i$ is the ith column of $\mathbf{V}_n$,
      ii. or as this basic ROM augmented by $\mathbf{P}^*\nabla J(\mathbf{s}^{(\ell)})$ which gives

$$\mathbf{V}^{(\ell)} = [(\mathbf{V}_n)_{\ell k+1}, \cdots, (\mathbf{V}_n)_{(\ell+1)k}, \mathbf{v}] \in \mathbb{R}^{n \times (k+1)}$$

        where $\mathbf{v}$ is defined by (16b) with $\mathbf{g} = \mathbf{P}^*\nabla J(\mathbf{s}^{(\ell)})$.
    c. Compute orthogonal projection on current subspace $\mathbf{V}^{(\ell)}$

$$\mathbf{Q} = \mathbf{V}^{(\ell)}(\mathbf{V}^{(\ell)})^T\mathbf{M}.$$

    d. For the Hessian approximation $\widehat{\mathbf{H}}^{(\ell)}$ obtained by $\mathbf{V}^{(\ell)}$, solve

$$\mathbf{Q}^*(\widehat{\mathbf{H}}^{(\ell)} + \beta\mathbf{I})\mathbf{Q}\Delta\mathbf{s} = -\mathbf{Q}^*\nabla J(\mathbf{s}^{(\ell)}) \tag{50}$$

e. Update $\mathbf{s}^{(\ell+1)} = \mathbf{s}^{(\ell)} + \mathbf{Q}\varDelta\mathbf{s}$.
f. Set $\mathbf{V}^{\text{old}} = [\mathbf{V}^{\text{old}}\ \mathbf{V}_1^{(\ell)}\ldots\mathbf{V}_k^{(\ell)}] \in \mathbb{R}^{n\times(\ell+1)k}$, where $\mathbf{V}_i^{(\ell)}$ is the ith column of $\mathbf{V}^{(\ell)}$.
g. Let $\mathbf{P} = \mathbf{I} - \mathbf{V}^{\text{old}}(\mathbf{V}^{\text{old}})^T\mathbf{M}$ be the $\mathbf{M}$-orthogonal projection on the complement of $\mathbf{V}^{\text{old}}$.

All comments in Remark 1 apply to Algorithm 6 as well.

As we have seen in subsection "Sequential Hessian Approximation Using the Basic ROM", of Sect. 4 there is a big difference between Algorithms 5 and 6 when the basic ROM is used. The gradient norm can increase from one iterate to the next if Algorithm 5 is used with the basic ROM (see (30)), whereas the gradient norm decreases monotonically if Algorithm 6 is used with the basic ROM (see (32)). However, if the augmented ROM is used, Algorithms 5 and 6 are identical, if the subsystems (47) and (50) are solved exactly, as we will show next.

Theorem 5 *Algorithms 5 and 6 with augmented ROMs are identical.*

Proof In step ℓ, assume that both algorithms have the same current iterate $\mathbf{s}^{(\ell)}$. Using $\mathbf{P}^* = \mathbf{P}$, $\mathbf{Q}^* = \mathbf{Q}$, the linear systems (47) and (50) read

$$\mathbf{P}(\widehat{\mathbf{H}}^{(\ell)} + \beta\mathbf{I})\mathbf{P}\varDelta\mathbf{s}_1 = -\mathbf{P}\nabla J(\mathbf{s}^{(\ell)}), \tag{51a}$$

$$\mathbf{Q}(\widehat{\mathbf{H}}^{(\ell)} + \beta\mathbf{I})\mathbf{Q}\varDelta\mathbf{s}_2 = -\mathbf{Q}\nabla J(\mathbf{s}^{(\ell)}), \tag{51b}$$

and these linear systems have unique solutions $\mathbf{P}\varDelta\mathbf{s}_1$ and $\mathbf{Q}\varDelta\mathbf{s}_2$, respectively.

By construction of the augmented ROM,

$$\mathbf{P}\nabla J(\mathbf{s}^{(\ell)}) \in \text{range}(\mathbf{V}^{(\ell)}) = \text{range}(\mathbf{Q}).$$

Together with $\mathbf{QP} = \mathbf{PQ} = \mathbf{Q}$, this implies that (51a) can be written equivalently as

$$\mathbf{P}(\widehat{\mathbf{H}}^{(\ell)} + \beta\mathbf{I})\mathbf{P}\varDelta\mathbf{s}_1 = -\mathbf{Q}\nabla J(\mathbf{s}^{(\ell)}). \tag{52}$$

It is shown in [8, Eq. (2.14)] that

$$\widehat{\mathbf{H}}^{(\ell)} = \mathbf{V}^{(\ell)}\exp(-2\widehat{\mathbf{A}}^T\ T)\ (\mathbf{V}^{(\ell)})^T\mathbf{M}$$
$$+ \int_0^T \mathbf{V}^{(\ell)}\exp(-2\widehat{\mathbf{A}}^T\ \tau)(\mathbf{V}^{(\ell)})^T\mathbf{M}\,d\tau, \tag{53}$$

where $\widehat{\mathbf{A}} = (\mathbf{V}^{(\ell)})^T\mathbf{AV}^{(\ell)}$. (When comparing with [8, Eq. (2.14)] recall the assumptions (3) and also note that $\widehat{\mathbf{M}} = (\mathbf{V}^{(\ell)})^T\mathbf{MV}^{(\ell)} = \mathbf{I} \in \mathbb{R}^{(k+1)\times(k+1)}$). Using (53), $(\mathbf{V}^{\text{old}})^T\mathbf{MV}^{(\ell)} = \mathbf{0}$ and the definitions of $\mathbf{P}$, $\mathbf{Q}$ we obtain

$$\widehat{\mathbf{H}}^{(\ell)} = \mathbf{P}\widehat{\mathbf{H}}^{(\ell)}\mathbf{P} = \mathbf{Q}\widehat{\mathbf{H}}^{(\ell)}\mathbf{Q}.$$

Inserting these equations into (52) and using $\mathbf{QP} = \mathbf{PQ} = \mathbf{Q}$ gives

$$\beta \mathbf{P} \Delta \mathbf{s}_1 = -\mathbf{Q} \nabla J(\mathbf{s}^{(\ell)}) - \mathbf{Q} \widehat{\mathbf{H}}^{(\ell)} \mathbf{Q} \Delta \mathbf{s}_1 \text{ and}$$
$$\beta \mathbf{Q} \Delta \mathbf{s}_1 = \beta \mathbf{QP} \Delta \mathbf{s}_1 = -\mathbf{Q} \nabla J(\mathbf{s}^{(\ell)}) - \mathbf{Q} \widehat{\mathbf{H}}^{(\ell)} \mathbf{Q} \Delta \mathbf{s}_1,$$

which implies

$$\mathbf{P} \Delta \mathbf{s}_1 = \mathbf{Q} \Delta \mathbf{s}_1.$$

Inserting this identity into (51a), multiplying the resulting identity from the left by $\mathbf{Q}$ and using $\mathbf{QP} = \mathbf{Q}$ gives

$$\mathbf{Q}(\widehat{\mathbf{H}}^{(\ell)} + \beta \mathbf{I}) \mathbf{Q} \Delta \mathbf{s}_1 = -\mathbf{Q} \nabla J(\mathbf{s}^{(\ell)}).$$

Since (51b) has a unique solution,

$$\mathbf{Q} \Delta \mathbf{s}_2 = \mathbf{Q} \Delta \mathbf{s}_1 = \mathbf{P} \Delta \mathbf{s}_1,$$

and both Algorithms 5 and 6 generate the same new iterate $\mathbf{s}^{(\ell+1)}$. □

6 Numerical Results

We consider the following model problem. Let $\Omega = (0, 1)^3$ be the spatial domain and $T = 1$. Given scalars $\kappa > 0$, $\beta > 0$, $\gamma \in \mathbb{R}$, and functions $y_d \in L^2(\Omega \times (0, T))$, $f \in L^2(\Omega \times (0, T))$, the optimal control problem is

$$\min_{s \in L^2(\Omega)} \frac{1}{2} \int_0^T \int_\Omega (y(x, t) - y_d(x, t))^2 dx dt$$
$$+ \frac{1}{2} \int_\Omega (y(x, T) - y_d(x, T))^2 dx + \frac{\beta}{2} \int_\Omega s^2(x) dx, \quad (54a)$$

where for given $s \in L^2(\Omega)$, the function $y \in W(0, T)$ is the solution of

$$\frac{\partial y(x, t)}{\partial t} - \kappa \Delta y(x, t) + \gamma y(x, t) = f(x, t), \qquad x \in \Omega, t \in (0, T), \quad (54b)$$
$$y(x, t) = 0, \qquad x \in \partial \Omega, t \in (0, T), \quad (54c)$$
$$y(x, 0) = s(x), \qquad x \in \Omega. \quad (54d)$$

We set

$$\kappa = 0.1, \gamma = 0.5, \text{ and } \beta = 10^{-4}.$$

The desired/observed state y_d is the solution of (54b–d) with

$$s(x) = 2e^{-10\|x-x_1\|_2^2} + e^{-5\|x-x_2\|_2^2} + 2e^{-50\|x-x_3\|_2^2}$$
$$+ 2e^{-40\|x-x_4\|_2^2} + e^{-100\|x-x_5\|_2^2}, \tag{55}$$

where $x_1 = (0.2, 0.2, 0.2), x_2 = (0.8, 0.8, 0.8), x_3 = (0.5, 0.5, 0.5), x_4 = (0.2, 0.8, 0.8), x_5 = (0.8, 0.2, 0.2)$.

We discretize (54) using P1 piecewise linear finite elements in space. More precisely, we divide $\Omega = (0, 1)^3$ into $n_{x_1} \times n_{x_2} \times n_{x_3} = 30 \times 30 \times 30$ cubes of equal size and then divide each cube into six tetrahedra. This leads to a problem of the type (1) of size $n = 29,791$ and with positive definite $\mathbf{A}$. For the time integration, we use the implicit Euler method with $n_t = 50$ steps.

We use basic ROMs of size $k = 20$. The initial iterate in all algorithms is chosen to be $\mathbf{s}^{(0)} = \mathbf{0}$, and the termination tolerance is set to tol $= 10^{-4}\|\nabla J(\mathbf{s}^{(0)})\|_{\mathbf{M}}$. In addition, we limit the maximum number of iterations in Algorithms 5 and 6 to $\ell_{\max} = 15$, which requires computation of the 300 smallest generalized eigenvalues of $(\mathbf{A}, \mathbf{M})$. These were computed using `eigs` in Matlab. The linear systems inside Algorithms 5 and 6 are solved using the CG method with residual tolerance given by (48) with $c_{\text{safe}} = 0.9$.

Table 1 below shows the convergence of the CG algorithm applied to the full-order problem, as well as the convergence behaviors of Algorithms 5 and 6 both with basic and with augmented ROM. 'Full solves' counts how many full-order state or adjoint-type Eqs. ((1c,d), (5), (7) and (8)) are solved, whereas 'ROM solves' counts how many reduced-order state or adjoint type Eqs. (12) and (13) are solved. In all cases, a function $J(\mathbf{s}^{(\ell)})$ and gradient $\nabla J(\mathbf{s}^{(\ell)})$ evaluation combined requires two 'full solves,' a Hessian–vector-multiplication $(\mathbf{H} + \beta\mathbf{I})\mathbf{s}$ requires two 'full solves' (7) and (8), whereas a ROM Hessian–vector-multiplication $(\widehat{\mathbf{H}}^{(\ell)} + \beta\mathbf{I})\mathbf{s}$ requires two 'ROM solves' (12) and (13). The number of 'full solves' is a good indicator of the overall computational expense of the algorithm.

In particular, Table 1 shows that Algorithm 5 with basic ROM generates iterates with increasing gradient norms in the first steps (it will generate an approximate solution if sufficiently more steps are executed). This behavior is explained by Theorem 1 and the inequality (30). In fact, Fig. 1 shows that $h(\lambda_i) = \widehat{\mathbf{H}}_{ii} > \beta$ for all i, which means that inequality (30) is valid for all steps ℓ. We stop Algorithm 5 with basic ROM after $\ell = 5$ steps. Algorithm 6 with basic ROM stops after $\ell_{\max} = 15$ iterations, but the gradient norm $\|\nabla J(\mathbf{s}^{(15)})\|_{\mathbf{M}} \approx 1.2 \, 10^{-4}$ is still relative large. The table also shows that when the augmented ROM is used, Algorithms 5 and 6 behave identically, which illustrates Theorem 5. More importantly, the augmented ROM significantly improves the performance of these algorithms over the basic ROM, and over the applying the CG algorithm directly to the full problem.

For this example, to reach the tolerance tol $= 10^{-4}\|\nabla J(\mathbf{s}^{(0)})\|_{\mathbf{M}}$, CG applied to the full-order problem requires 50 full-order PDE solves, Algorithm 6 with basic ROM requires more than 30 full-order PDE solves, whereas Algorithm 6 with aug-

Table 1 Convergence of the CG algorithm applied to the full-order problem, and of Algorithms 5 and 6 both with basic and with augmented ROM. Algorithm 6 with augmented ROM is the most efficient one and only requires 20 full-order model solves compared to the 50 full-order model solves required by CG applied to the original problem

		iter	J	$\|\nabla J\|$	CG iters	full/ROM solves
	Full Hessian	0	1.4578e-02	5.9128e-02		2/0
		1	1.7529e-05	5.6103e-06	24	48/0
Algorithm 5	Basic ROM	0	1.4578e-02	5.9128e-02		2/0
		1	1.8157e+01	8.4731e-01	13	2/26
		2	4.0922e+04	3.3553e+01	11	2/22
		3	1.9961e+07	6.3194e+02	10	2/20
		4	3.7777e+10	2.4762e+04	11	2/22
		5	2.6355e+13	6.1620e+05	11	2/22
	Augmented ROM	0	1.4578e-02	5.9128e-02		2/0
		1	5.2692e-05	8.6795e-04	12	2/24
		2	2.4754e-05	3.3800e-04	7	2/14
		3	1.9291e-05	1.5485e-04	6	2/12
		4	1.8066e-05	6.7495e-05	5	2/10
		5	1.7729e-05	3.1675e-05	5	2/10
		6	1.7693e-05	3.5444e-05	5	2/10
		7	1.7634e-05	3.1536e-05	5	2/10
		8	1.7542e-05	6.0700e-06	4	2/8
		9	1.7527e-05	2.4308e-06	4	2/8
Algorithm 6	Basic ROM	0	1.4578e-02	5.9128e-02		2/0
		1	5.5269e-04	4.3797e-03	11	2/22
		2	1.7462e-04	1.9109e-03	7	2/14
		3	6.9784e-05	8.5360e-04	5	2/10
		4	6.0665e-05	7.2887e-04	5	2/10
		5	4.4527e-05	5.1389e-04	4	2/8
		6	4.0420e-05	4.5001e-04	4	2/8
		7	3.3630e-05	3.4294e-04	4	2/8
		8	3.3074e-05	3.3361e-04	4	2/8
		9	2.8848e-05	2.6046e-04	4	2/8
		10	2.8542e-05	2.5511e-04	4	2/8
		11	2.6371e-05	2.1818e-04	4	2/8
		12	2.4632e-05	1.8407e-04	4	2/8
		13	2.4513e-05	1.8171e-04	4	2/8
		14	2.3966e-05	1.7130e-04	3	2/6
		15	2.1713e-05	1.2443e-04	4	2/8

(continued)

Table 1 (continued)

	iter	J	$\|\nabla J\|$	CG iters	full/ROM solves
Augmented ROM	0	1.4578e-02	5.9128e-02		2/0
	1	5.2692e-05	8.6795e-04	12	2/24
	2	2.4754e-05	3.3800e-04	7	2/14
	3	1.9291e-05	1.5485e-04	6	2/12
	4	1.8066e-05	6.7495e-05	5	2/10
	5	1.7729e-05	3.1675e-05	5	2/10
	6	1.7693e-05	3.5444e-05	5	2/10
	7	1.7634e-05	3.1536e-05	5	2/10
	8	1.7542e-05	6.0700e-06	4	2/8
	9	1.7527e-05	2.4308e-06	4	2/8

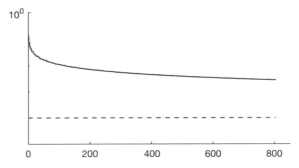

Fig. 1 The 800 largest Hessian entries $h(\lambda_i) = \widetilde{\mathbf{H}}_{ii}$ (solid line) for the diagonalized problem and β (dashed line). The Hessian entries are larger than β which means that the basic ROM leads to less good Hessian approximations

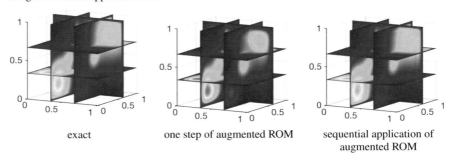

| exact | one step of augmented ROM | sequential application of augmented ROM |

Fig. 2 Exact initial state (55) (left), initial state computed using one step of the augmented ROM (middle) and initial state $s^{(9)}$ computed using sequential application of the augmented ROM (right). While one step of the augmented ROM gives a good approximation of the exact initial state, the sequential application of the augmented ROM substantially improves this approximation

mented ROM requires only 20 full-order PDE solves. Thus, the cost of our sequential augmented ROM is about 40% the cost of using the full-order model.

Figure 2 shows the exact initial state (55), the initial state computed using one step of the augmented ROM (Algorithm 6 with augmented ROM and $\ell_{max} = 0$, which is our previous approach [8]), and the initial state computed using Algorithm 6 with augmented ROM). While one step of the augmented ROM already gives a good approximation of the exact initial state, the initial state $s^{(9)}$ computed with the sequential application of the augmented ROM is in excellent agreement with the exact initial state (which is expected since $\|\nabla J(s^{(9)})\| \approx 2.4 \cdot 10^{-6}$).

7 Conclusions

We have developed a rigorous approach for the sequential application of the augmented ROM introduced in [8]. The proposed sequential application of the augmented ROM can compute an approximate control with the same accuracy as the one obtained using the expensive full-order model, but at a fraction of the cost (about 40% in the numerical example shown). Our sequential application of the augmented ROM applies to problems that satisfy (3). The extension of this sequential approximation to a broader class of problems, especially problems with nonsymmetric $\mathbf{A}$, is part of future work.

Acknowledgements This work was supported in part by NSF grant DMS-1522798, by the DARPA EQUiPS Program, Award UTA15-001068, and by the Excellence Initiative of the German Federal Government and the state governments—funding line Institutional Strategy: DFG project number ZUK 49/2 ("Heidelberg: Realizing the Potential of a Comprehensive University")—via the Heidelberg Mobility Programme.

References

1. Akçelik, V., Biros, G., Ghattas, O., Long, K.R., van Bloemen Waanders, B.: A variational finite element method for source inversion for convective-diffusive transport. Finite Elem. Anal. Des. **39**(8), 683–705 (2003). https://doi.org/10.1016/S0168-874X(03)00054-4
2. Bashir, O., Willcox, K., Ghattas, O., van Bloemen Waanders, B., Hill, J.: Hessian-based model reduction for large-scale systems with initial-condition inputs. Int. J. Numer. Methods Eng. **73**(6), 844–868 (2008). https://doi.org/10.1002/nme.2100
3. Benner, P., Sachs, E., Volkwein, S.: Model order reduction for PDE constrained optimization. In: Leugering, G., Benner, P., Engell, S., Griewank, A., Harbrecht, H., Hinze, M., Rannacher, R., Ulbrich, S. (eds.) Trends in PDE Constrained Optimization. International Series of Numerical Mathematics, vol. 165, pp. 303–326. Birkhäuser/Springer, Cham (2014). http://dx.doi.org/10.1007/978-3-319-05083-6_19
4. Blum, J., Le Dimet, F.X., Navon, I.M.: Data assimilation for geophysical fluids. In: Temam, R.M., Tribbia, J.J. (eds.) Handbook of Numerical Analysis, vol. XIV. Special volume: Computational Methods for the Atmosphere and the Oceans, vol. 14, pp. 385–441. Elsevier/North-Holland, Amsterdam (2009). http://dx.doi.org/10.1016/S1570-8659(08)00209-3

5. Carraro, T., Geiger, M.: Direct and indirect multiple shooting for parabolic optimal control problems. In: Carraro, T., Geiger, M., Körkel, S., Rannacher, R. (eds.) Multiple Shooting and Time Domain Decomposition Methods. MuS-TDD, Heidelberg, 6-8 May 2013. Contributions in Mathematical and Computational Sciences, vol. 9, pp. 35–67. Springer, Heidelberg (2015). https://doi.org/10.1007/978-3-319-23321-5_2

6. Daescu, D.N., Navon, I.M.: Efficiency of a POD-based reduced second-order adjoint model in 4d-var data assimilation. Int. J. Numer. Methods Fluids 53(6), 985–1004 (2007). http://dx.doi.org/10.1002/fld.1316

7. Gubisch, M., Volkwein, S.: Proper orthogonal decomposition for linear-quadratic optimal control. In: Benner, P., Cohen, A., Ohlberger, M., Willcox, K. (eds.) Model Reduction and Approximation: Theory and Algorithms, Computational Science and Engineering, pp. 3–64. SIAM, Philadelphia (2017). https://doi.org/10.1137/1.9781611974829.ch1

8. Heinkenschloss, M., Jando, D.: Reduced order modeling for time-dependent optimization problems with initial value controls. SIAM J. Sci. Comput. 40(1), A22–A51 (2018). https://doi.org/10.1137/16M1109084

9. Hesse, H.K., Kanschat, G.: Mesh adaptive multiple shooting for partial differential equations. I. Linear quadratic optimal control problems. J. Numer. Math. 17(3), 195–217 (2009). http://dx.doi.org/10.1515/JNUM.2009.011

10. Hinze, M., Pinnau, R., Ulbrich, M., Ulbrich, S.: Optimization with PDE Constraints. Mathematical Modelling, Theory and Applications, vol. 23. Springer, Heidelberg (2009). http://dx.doi.org/10.1007/978-1-4020-8839-1

11. Lehoucq, R.B., Sorensen, D.C., Yang, C.: ARPACK User's Guide: Solution of large Scale Eigenvalue Problems by Implicitly Restarted Arnoldi Methods. SIAM, Philadelphia (1998). https://doi.org/10.1137/1.9780898719628

12. Liberzon, D.: Calculus of Variations and Optimal Control Theory. A Concise Introduction. Princeton University Press, Princeton (2012)

13. Rao, V., Sandu, A.: A time-parallel approach to strong-constraint four-dimensional variational data assimilation. J. Comput. Phys. 313, 583–593 (2016). http://dx.doi.org/10.1016/j.jcp.2016.02.040

14. Sachs, E.W., Volkwein, S.: POD-Galerkin approximations in PDE-constrained optimization. GAMM-Mitteilungen 33(2), 194–208 (2010). http://dx.doi.org/10.1002/gamm.201010015

A Direct Index 1 DAE Model of Gas Networks

Peter Benner, Maike Braukmüller and Sara Grundel

Abstract Using isothermal Euler equations and a network graph to model gas flow in a pipeline network is a classical description, and we prove that any direct space discretization results in a system of index 2 nonlinear differential algebraic equations (DAE). Those are hard to simulate, and model order reduction techniques are not very developed for this system class. However, we can show that a simple approximation results in an index 1 system of nonlinear differential algebraic equations, which is easier to simulate and we can show that a structured projection leads to a reduced system that also typically has index 1. We validate the use of this model and its advantage for fast simulation, including model order reduction, in some numerical examples.

1 Introduction

In the past years, many investigations on the efficient simulation of complex gas networks were made [7–10, 13, 25]; however, there are still many open challenges, especially in the field of applied mathematics [5]. First of all, the modeling of the gas flow within each pipe leads to partial differential equations. On top of that, there are additional network elements, e.g., resistors and valves, increasing the complexity of the representing model. The physics, on the one hand, and the network structure, on the other hand, need to be combined, yielding a system of (nonlinear) partial differential algebraic equations. In general, it is hard to simulate such systems within

P. Benner · S. Grundel (✉)
Max Planck Institute for Dynamics of Complex Technical Systems, Sandtorstr. 1, 39016
Magdeburg, Germany
e-mail: grundel@mpi-magdeburg.mpg.de

P. Benner
e-mail: benner@mpi-magdeburg.mpg.de

M. Braukmüller
Institute Computational Mathematics TU Braunschweig, Universitätsplatz 2, 31806
Braunschweig, Germany
e-mail: MaikeBraukmueller@gmx.de

© Springer International Publishing AG, part of Springer Nature 2018
W. Keiper et al. (eds.), *Reduced-Order Modeling (ROM) for Simulation and Optimization*,
https://doi.org/10.1007/978-3-319-75319-5_5

a reasonable computation time. In particular, once the size of the system becomes large we would like to use model order reduction techniques to create truly efficient simulation tools for operation and control of the network. The aim of model order reduction is to approximate large-scale problems, e.g., arising from gas transportation networks, with lower dimensional models, keeping the loss of accuracy as small as possible in order to lower the computation time. In other words, model order reduction tries to find a compromise between speeding up the simulation and keeping the error small. This paper looks into the most direct and straightforward way that one could discretize such a network system. Even though we will use a simple finite difference discretization scheme, the resulting structure of the network is similar if one uses more interesting and accurate discretization methods for each individual pipe. We will, however, show that this approach will lead to a differential algebraic equation (DAE) of index 2. The index of a DAE is a measure of how far away the equation is from an ordinary differential equation (ODE). The higher the index is the harder it is to simulate such a system. One definition of an index will be given in the paper. In general, we can say that a DAE of index 1, which is almost an ODE, is typically easy to solve and some standard ODE solvers, including some of the MATLAB® solvers, are able to do so. This is why we want to stress that using a simple approximation of the isothermal Euler equation, which is the underlying partial differential equation, given by [13] results in a DAE of index 1. For simulation purposes, index 1 DAEs are almost as good as ODEs, however, for model order reduction that is not entirely true, as there is no guarantee that projection of a DAE of index 1 will result in a reduced DAE of index 1 as well. One can show that the index can increase in general. However, we will investigate what happens if we use a structured projection and we will see that typically the reduced model is also of index 1. Since the equations are nonlinear and we need to create structured projection matrices, we decided to use Proper Orthogonal Decomposition (POD) as a model order reduction method to show that it is possible to reduce these systems and that a speedup can be achieved in this case even without using hyperreduction methods. POD is a method well studied in the model order reduction community, and the applications are manifold [3, 6, 12, 16, 22].

This paper takes the standard modeling of gas pipeline networks and considers an arbitrary fine simple space discretization, which is discussed in detail in Sect. 2. An approximation based on [13] is also introduced there. Given that the resulting set of equations is a system of differential algebraic equations, we investigate their index, which is proved to be typically 2 for the given simple discretization scheme and at least 2 for any space discretization scheme; see Sect. 3. Furthermore, it is proved that the DAE, which is based on the approximation in [13], results in a DAE of index 1. In Sect. 4, we describe the standard model order reduction method Proper Orthogonal Decomposition (POD) and how we use it for our system to keep the structure. The numerical results (Sect. 5) show that the approximation is accurate and already improves the computation time. We also investigate the use of POD to create a further speedup.

2 The Model

In order to model a gas transportation pipeline network, we need to describe the physics of the gas transportation in one pipe and how the pipes are connected in the network, for which we use a graph. The physics is described by a partial differential equation. Discretizing this in space leads to an ordinary differential equation, but in combination with the algebraic constraint given by putting the pipes together over the graph structure, the resulting system is a differential algebraic equation. In the following, we will derive this differential algebraic equation by using one particular spatial discretization scheme.

Gas Physics

We base our basic model on the isothermal Euler equation with friction-dominated term. We neglect the term for the kinetic energy, and we use a simple linear dependency between pressure and pressure density. This is given by the following set of equations:

$$\frac{\partial}{\partial t}\rho(t, x) + \frac{\partial}{\partial x}q(t, x) = 0, \tag{1a}$$

$$\frac{\partial}{\partial t}q(t, x) + \frac{\partial}{\partial x}p(t, x) = -\lambda\frac{q(t, x)|q(t, x)|}{2D\rho(t, x)}, \tag{1b}$$

$$p(t, x) = c_k\rho(t, x), \tag{1c}$$

with constant friction factor λ, density $\rho(t, x)$, pressure $p(x, t)$, and flux in the pipe $q(t, x)$. Other parameters are the diameter D and sound speed, expressed by the constant c_k. The pressure density is measured in kg/m^3, the pressure in Pascal (kg/(ms^2)), and the flux in kg/(m^{2}s).

Equation (1a) is called the continuity equation and Eq. (1b) the momentum equation. For a detailed description and derivation of the isothermal Euler equations, we refer to [21]. Equation (1c) is the state equation of a real gas. It relates the density $\rho(t, x)$ and the pressure $p(t, x)$. In general, the connection between pressure and density is more complex than stated above. Equation (1c), the ideal gas equation, is an approximation and, as the name indicates, only holds for ideal gases.

The first two equations are the so-called Weymouth equations. We substitute the pressure by the pressure density to get a simple transport partial differential equation that approximates the gas transportation in a pipe. We furthermore replace the flow by the mass flow, which is the flow multiplied with the pipe's cross section $a = \pi/4D^2$ yielding

$$\frac{\partial}{\partial t}\rho(t, x) + \frac{\partial}{\partial x}\frac{1}{a}q(t, x) = 0, \tag{2a}$$

$$\frac{\partial}{\partial t}q(t, x) + a\frac{\partial}{\partial x}c_k\rho(t, x) = -\lambda\frac{q(t, x)|q(t, x)|}{2Da\rho(t, x)}, \tag{2b}$$

where $q(x, t)$ is no longer the flux but the mass flux.

The Network as a Graph

It is common to model gas networks as oriented graphs; see [14]. Let $G = (\mathcal{N}, \mathcal{E})$ be an oriented graph with a set of nodes, $\mathcal{N} = \{n_1, n_2, ..., n_N\}$, and a set of branches, $\mathcal{E} = \{e_1, e_2, ..., e_E\}$. The nodes represent pipe junctions or sources and sinks in the network. The branches represent the pipes of the network and could also refer to other elements, like valves or resistors. However, in this paper networks consisting of pipes only will be considered. Each branch e_j, $j = 1, ..., E$, is directed from its left node $(n_L)_j$ to its right node $(n_R)_j$.

With incidence matrices, $A, A_R, A_L \in \mathbb{R}^{N \times E}$, we can express the relationship between the nodes and pipes. The matrices A_L and A_R indicate the left and right nodes of each branch, respectively, and are defined as

$$(A_L)_{ij} = \begin{cases} -1 & n_i \text{ is left node of branch } e_j, \\ 0 & \text{else,} \end{cases}$$

$$(A_R)_{ij} = \begin{cases} +1 & n_i \text{ is right node of branch } e_j, \\ 0 & \text{else.} \end{cases}$$

The overall network topology is expressed by the matrix $A := A_L + A_R$,

$$(A)_{ij} = \begin{cases} -1 & n_i \text{ is left node of branch } e_j, \\ +1 & n_i \text{ is right node of branch } e_j, \\ 0 & \text{else,} \end{cases}$$

$i = 1, ..., N$, $j = 1, ..., E$.

To ensure the correctness of the definition of A, we need to exclude self-loops, meaning branches that have the same node as their left and right nodes. The gas flow is described by the pressure density and the mass flow within the network. For each pipe e_j, we have the mass flow $q_j(x, t)$ and the density $\rho_j(x, t)$, with $x \in [0, L_j]$, where L_j is the length of the pipe. We also introduce $\rho^i(t)$ as the pressure density at node n_i. The direction of the pipes given by the graph topology is not necessarily the direction of the flow. Hence, we obtain the flow direction by interpreting the sign of the flow variables. A positive sign means that the gas flows from the left to the right node; a negative sign means that it flows from the right to the left node. Our networks consist of sources and sinks and junctions. At sources, the pressure density is given, at sinks the mass flow is given, and at junctions we require the sum of all mass flows to be equal to zero. Let $\mathcal{N}_0$ be junctions, $\mathcal{N}_s$ be sources or supply nodes, and $\mathcal{N}_d$ be sinks or demand nodes. The set $\mathcal{E}_i^L = \{e \in \mathcal{E} : n_i \text{ is left node of } e\} \subseteq \mathcal{E}$, containing all edges which have n_i as their left node, and analogously $\mathcal{E}_i^R = \{e \in \mathcal{E} : n_i \text{ is right node of } e\} \subseteq \mathcal{E}$, containing all edges with n_i being their right node, are used to define the network conditions:

$$\rho^i(t) = s^i(t) \qquad\qquad\qquad \text{for } n_i \in \mathcal{N}_s, \qquad (3a)$$

$$\sum_{e_j \in \mathcal{E}_i^R} q_j(t, L_j) - \sum_{e_j \in \mathcal{E}_i^L} q_j(t, 0) = 0 \qquad\qquad \text{for } n_i \in \mathcal{N}_0, \qquad (3b)$$

$$\sum_{e_j \in \mathcal{E}_i^R} q_j(t, L_j) - \sum_{e_j \in \mathcal{E}_i^L} q_j(t, 0) = d^i(t) \qquad\qquad \text{for } n_i \in \mathcal{N}_d. \qquad (3c)$$

Here, $s^i(t)$ is the given pressure density at the node n_i and $d^i(t)$ the given mass flux at node n_i. Using the incidence matrices A_L^0 and A_R^0 which are the matrices A_L and A_R with the columns of the supply nodes removed, we can write Eqs. (3b) and (3c) in matrix form as

$$A_L^0 Q_L(t) + A_R^0 Q_R(t) = d(t),$$

where the vector Q_L contains all the left fluxes and Q_R all the right fluxes of each pipe and the vector d contains the demand functions and zeros accordingly.

Discretization

Assuming we have a given network of gas pipes, represented by the graph structure, as above, and assuming the set of equations in (2) is given on every pipe, we will now use a simple space discretization scheme on each pipe partial differential equation. We discretize the pipe e_j into m_j segments, shown in Fig. 1, creating $m_j - 1$ interior discretization points. We define a mass flux on each segment and a pressure density on each node, denoted by $Q_1^j, \ldots, Q_{m_j}^j$ and $P_0^j, \ldots, P_{m_j}^j$. The left pressure P_L^j is equal to P_0^j and similar on the right. This means that for each pipe j we get a set of ordinary differential equations given by:

$$\frac{\partial}{\partial t} P_l^j + \left(\frac{Q_{l+1}^j - Q_l^j}{a_j h^j} \right) = 0, \qquad\qquad l = 1, \ldots, m^j - 1, \qquad (4a)$$

$$\frac{\partial}{\partial t} Q_l^j + c_k a_j \left(\frac{P_{l+1}^j - P_l^j}{h^j} \right) = -\lambda \frac{Q_l^j |Q_l^j|}{2D P_l^j}, \qquad\qquad l = 1, \ldots, m^j. \qquad (4b)$$

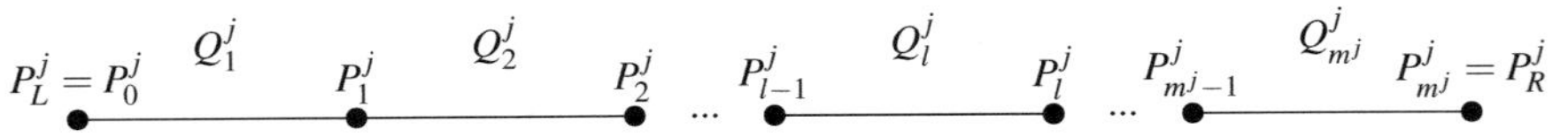

$$P_L^j = P_0^j \quad\overset{Q_1^j}{\rule{2em}{0pt}}\quad P_1^j \quad\overset{Q_2^j}{\rule{2em}{0pt}}\quad P_2^j \quad \cdots \quad P_{l-1}^j \quad\overset{Q_l^j}{\rule{2em}{0pt}}\quad P_l^j \quad \cdots \quad P_{m^j-1}^j \quad\overset{Q_{m^j}^j}{\rule{2em}{0pt}}\quad P_{m^j}^j = P_R^j$$

Fig. 1 Pipe discretization for pipe e_j

Let us now define the vector P to be a long vector containing the interior pressure nodes of each pipe, Q to be the vector containing all flux variables for all pipes, and p to be the vector of pressures at the original nodes that are not supply nodes. Let D_1 and D_2 be block diagonal matrices that describe the simple space discretization, then the gas pipeline system reads

$$\partial_t P = D_1 Q, \tag{5a}$$
$$\partial_t Q = D_2 P + Bp + f(P, Q, p, s(t)), \tag{5b}$$
$$0 = CQ - d(t). \tag{5c}$$

The matrix B is needed to pick up the left and right boundary pressures that are stored in p, and the nonlinear function f describes the nonlinearity given in (4b) and also uses the given pressure density function $s^k(t)$ for the supply nodes. The matrix $C = A_L^0 B_L + A_R^0 B_R$ is a sum of two matrix products, where B_L and B_R are matrices picking the left and right fluxes of each pipe from the vector of all fluxes Q and A_L and A_R are the left and right incidence matrices.

Let $x = \begin{pmatrix} P \\ Q \\ p \end{pmatrix}$, then the equation can be written as

$$E\dot{x} = Ax + F(x, s(t), d(t)) \tag{6}$$

with

$$A = \begin{pmatrix} 0 & D_1 & 0 \\ D_2 & 0 & B \\ 0 & C & 0 \end{pmatrix}, \quad E = \begin{pmatrix} I & 0 & 0 \\ 0 & I & 0 \\ 0 & 0 & 0 \end{pmatrix}, \quad F(x, t) = \begin{pmatrix} 0 \\ f(P, Q, p, s(t)) \\ d(t) \end{pmatrix}.$$

This DAE will be the basis for our discussion. The description depends on the given network with all the parameters and on the space discretization. The solution depends on that and the input functions $s(t)$ and $d(t)$ as well as the starting configuration $x(0)$. Eq. (1) neglects the kinetic energy, since it is several orders of magnitude smaller than the dominating terms. In [13], Herty et al. show that the time derivative of the flux is also several orders of magnitude smaller. They neglect the term as well and show convincing numerical results for a one pipe system. We use this approximation to get the following DAE:

$$\partial_t P = D_1 Q, \tag{7a}$$

$$0 = D_2 P + Bp + f(P, Q, p, s(t)), \tag{7b}$$

$$0 = CQ + d(t). \tag{7c}$$

These two systems—the DAE given in (5) and the DAE given in (7)—will be considered in the following and compared in terms of their index, their simulation properties, and their behavior for model order reduction.

3   Index

The systems (5) and (7) are both DAEs. A DAE is typically characterized by its index. Many different index concepts exist, and we will use the notion of the tractability index. In this section, we define the tractability index and show that the first DAE has tractability index 2 and the second has tractability index 1, which makes the second description more desirable. Depending on the focus, an index gives information on the difficulty of dealing with analytical or numerical aspects of the DAE [19]. The tractability index introduced in [18] for linear DAEs and extended for nonlinear DAEs, in [17], is a proper choice for the given problem [11].

Tractability Index

Let us consider a nonlinear DAE of the form (6). To define the tractability index, consider matrix sequence G_i, $i = 0, 1, 2, ...$, given by

$$G_0 = E, \tag{8a}$$

$$G_i = G_{i-1} + B_{i-1} Q_{i-1}, \tag{8b}$$

$$B_0 = A + J_x F, \tag{8c}$$

$$B_i = B_{i-1} P_{i-1}, \tag{8d}$$

$$P_i = I - Q_i, \tag{8e}$$

with Q_i being a projector onto ker G_i and $J_x F$ the Jacobian matrix of F at x. The following definitions can be found in [17].

Definition 1 The sequence (8) is called an *admissible matrix function sequence* if the matrix G_i has constant rank r_i, $i \in \mathbb{N}$.
The DAE system (4) is said to be

1. *regular with tractability index* 0, if G_0 has full rank;
2. *regular with tractability index* μ, if the associated matrix sequence is admissible, rank$(G_{\mu-1})$ < rank(G_μ), and if G_μ has full rank;

3. *regular*, if it is regular with any index $\mu \in \mathbb{N}$.

Based on this, we can now prove the following:

Lemma 2 *The tractability index of the DAE given in (5) is at least 2.*

Proof Following the description of (8), we get that

$$G_0 = E = \begin{pmatrix} 1 & 0 & 0 \\ 0 & 1 & 0 \\ 0 & 0 & 0 \end{pmatrix}, \quad B_0 = \begin{pmatrix} 0 & D_1 & 0 \\ D_2 + J_P f & J_Q f & B + J_p f \\ 0 & C & 0 \end{pmatrix} \tag{9}$$

With

$$Q_0 = \begin{pmatrix} 0 & 0 & 0 \\ 0 & 0 & 0 \\ 0 & 0 & 1 \end{pmatrix} \tag{10}$$

we get

$$G_1 = G_0 + B_0 Q_0 = \begin{pmatrix} 1 & 0 & 0 \\ 0 & 1 & B + J_p f \\ 0 & 0 & 0 \end{pmatrix} \tag{11}$$

which does not have full, but constant rank. Therefore, the DAE has a tractability index of at least 2. A projector onto the kernel of G_1 is then given by

$$Q_1 = \begin{pmatrix} 0 & 0 & 0 \\ 0 & 0 & -B - J_p f \\ 0 & 0 & 1 \end{pmatrix}$$

and

$$B_1 = B_0(I - Q_0) = \begin{pmatrix} 0 & D_1 & 0 \\ D_2 + J_P f & J_Q f & 0 \\ 0 & C & 0 \end{pmatrix},$$

$$G_2 = G_1 + B_1 Q_1 = \begin{pmatrix} 1 & 0 & -D_1(B + J_p f) \\ 0 & 1 & -J_Q f(B + J_p f) \\ 0 & 0 & -C(B + J_p f) \end{pmatrix}.$$

This means that the tractability index is exactly 2 if $-C(B + J_p f)$ has full rank.

Theorem 3 *The tractability index of the DAE given in (7) is at least 1 and exactly 1 iff $J_Q f$ is invertible and $C J_Q f^{-1}(B + J_p f)$ is invertible.*

Proof (*Theorem* 3) Following the description of (8), we get that

$$G_0 = E = \begin{pmatrix} 1 & 0 & 0 \\ 0 & 0 & 0 \\ 0 & 0 & 0 \end{pmatrix}, \quad B_0 = \begin{pmatrix} 0 & D_1 & 0 \\ D_2 + J_P f & J_Q f & B + J_p f \\ 0 & C & 0 \end{pmatrix} \tag{12}$$

With

$$Q_0 = \begin{pmatrix} 0 & 0 & 0 \\ 0 & 1 & 0 \\ 0 & 0 & 1 \end{pmatrix}, \tag{13}$$

we get

$$G_1 = G_0 + B_0 Q_0 = \begin{pmatrix} 1 & D_1 & 0 \\ 0 & J_Q f & B + J_p f \\ 0 & C & 0 \end{pmatrix}, \tag{14}$$

which does have full rank if $J_Q f$ is invertible and $C J_Q f^{-1}(B + J_p f)$ has full rank, which proves the theorem.

$J_Q f$ is a diagonal matrix, and if the flux in the pipe is nonzero, the matrix is invertible. The Schur complement $C J_Q f^{-1}(B + J_p f)$ is invertible in the cases that we considered. If a DAE has index 1, even the MATLAB® standard ODE solver ode15s can integrate the system, whereas for a DAE with a higher index we have to use more specific solvers. This makes it very clear that it is of great advantage to work with an index 1 DAE.

Other Discretizations

Using other discretizations, the block structure of all the matrices stays the same. All zeros blocks stay zero. All the other blocks can be different and depend on the discretization itself. However, following the argument from above, the index can only be higher, which leads to the following corollary.

Corollary 4 *For any discretization scheme, the Weymouth equations over a pure pipeline network are of index at least 2. Removing the time derivative of the mass flux, the index is at least 1.*

4 Model Order Reduction

Real gas pipeline networks consist of several thousands of pipes, which can be several kilometers long. Once such a system is space discretized, the resulting DAE can be large, and therefore the simulation can be difficult and slow. Such problems are solved by using model order reduction (MOR). This is an active field of research but also an established one [2, 4, 24]. Modern techniques are developing that are especially designed for DAEs with higher index [1, 6, 23, 27], but we want to use a common and standard technique, namely POD which is good and powerful for ordinary nonlinear differential equations [12, 26]. In the following, we briefly repeat the concept of POD and how we used it to create reduced models of the approximated index 1 DAE description of the gas pipeline system.

Proper Orthogonal Decomposition

We will use POD as a method to create the reduced order model via projection. The POD will give us a matrix W with which we can project the full model to create a reduced model. Given a model

$$E\dot{x} = Ax + F(x, s(t), d(t))$$

as in (6), we create a reduced model as

$$W^T E W \dot{x}_r = W^T A W x_r + W^T F(W x_r, s(t), d(t)), \tag{15}$$

where the solution $x_r(t)$ can be lifted back such that hopefully $W x_r(t)$ approximates the solution $x(t)$ of (6) well. The basic idea of POD is to determine interdependencies between the entries of the state vector and to aggregate these in a vector of much smaller dimension. Thus, we hope to reduce the system to a small number of variables which are not related, in this case meaning orthogonal, to each other, [15].

In a continuous setting, POD gives a good approximation to the best linear subspace of a certain size approximating the solution space. POD is a Galerkin projection, and thus we need to find an orthogonal projection $\Pi = W W^T$. The aim for the projection $\Pi = W W^T$ is to minimize the error in the state space.

We introduce POD, also known as Karhunen–Loève decomposition or principal component analysis, similar to the explanations in [9, 11, 20]. Let $u_1, ..., u_r$ denote the orthonormal columns of W. The approximation of $x(t)$ can be written as the minimization problem

$$\min_{u_1,...,u_r} \int_0^T \left\| x(t) - \sum_{i=1}^r \langle x(t), u_i \rangle u_i \right\|^2 \, dt, \quad \langle u_i, u_j \rangle = \delta_{ij}. \tag{16}$$

Note that the norm could be any norm connected to an inner product; however, in our derivation we use the n-dimensional Euclidian norm $\| \cdot \|_2$,

$$\|x\|_2 = \left(\sum_{i=1}^{n} x_i^2\right)^{\frac{1}{2}},$$

and the related canonical inner product. In general, the continuous problem, Eq. (16), cannot be solved numerically. Due to that, we introduce the so-called method of snapshots. At distinct time steps $t_0, ..., t_N$, we compute the snapshots $x(t_0), ..., x(t_N)$ and define the snapshot matrix as $Y = [x(t_1), ..., x(t_N)]$. Approximating $x(t)$ with the snapshots, we obtain a discrete formulation for the POD minimization problem,

$$\min_{u_1,...,u_r} \sum_{k=1}^{N} \left\| x_k - \sum_{i=1}^{r} \langle x_k, u_i \rangle u_i \right\|_2^2, \quad \langle u_i, u_j \rangle = \delta_{ij}. \tag{17}$$

The solution of (17) is closely related to the singular value decomposition (SVD) of the snapshot matrix Y. Consider the SVD of Y,

$$Y = U \Sigma V^\top,$$

with orthogonal matrices $U \in \mathbb{R}^{n \times n}$, $V \in \mathbb{R}^{N \times N}$, singular values $\sigma_1 \geq ... \geq \sigma_d > 0$, and

$$\Sigma = \begin{pmatrix} \Sigma_d & 0 \\ 0 & 0 \end{pmatrix} \in \mathbb{R}^{n \times N}, \quad \Sigma_d = \text{diag}(\sigma_1, ..., \sigma_d).$$

The columns of $U = [u_1, ..., u_n]$ are called left singular vectors and the columns of $V = [v_1, ..., v_N]$ right singular vectors. The squared singular values σ_i^2 are the nonzero eigenvalues of $YY^\top$ and $Y^\top Y$, and the singular vectors are the associated eigenvectors. The vectors solving (17) are exactly the left singular vectors. Meaning the matrix W we are looking for contains the first r columns of the matrix U of the singular value decomposition of the snapshot matrix Y. One can do the same procedure using more than one time series creating a snapshot matrix that includes all snapshots of all scenarios.

Reduced Gas Models

Going back to our model which has the distinct block structure related to the three components of

$$x = \begin{pmatrix} P \\ Q \\ p \end{pmatrix},$$

we consider a structured and an unstructured version to create a reduced order model. It is in general possible to use any projection matrix to create a reduced model from the given DAE. However, the reduced model could be a DAE of any index, which is not desirable. We take the projection matrix from POD and snapshots of the x directly and call the reduced model generated from that the unstructured reduced model. We also use a structured projection matrix, meaning a projection matrix that is block diagonal such that the reduced system has the same block structure as the original system. This means that we create three snapshot matrices: one for the interior pressures Y_P, one for the fluxes Y_Q, and one for the node pressures Y_p. With that, we compute three projection matrices W_P, W_Q, W_p via POD and define the two reduced models according to (5) and (7) as

$$\partial_t P_r = W_P^T D_1 W_Q Q,$$
$$\partial_t Q_r = W_Q^T D_2 W_P P_r + W_Q^T B W_p p_r + W_Q^T f(W_P P_r, W_Q Q_r, W_p p_r, s(t)),$$
$$0 = W_p^T C W_Q Q_r + d(t).$$

and

$$\partial_t P_r = W_P^T D_1 W_Q Q, \tag{18a}$$
$$0 = W_Q^T D_2 W_P P_r + W_Q^T B W_p p_r + W_Q^T f(W_P P_r, W_Q Q_r, W_p p_r, s(t)), \tag{18b}$$
$$0 = W_p^T C W_Q Q_r + d(t). \tag{18c}$$

These are once more DAEs, and they obey the structure. This means that the reduced DAE of the standard model is at least of index 2, and the reduced second model is of index at least 1. Following the proof of Theorem 3, we can see that the reduced system (18) has index 1 exactly when $W_Q^T J_Q f W_Q$ and

$$W_p^T C W_Q (W_Q^T J_Q f W_Q)^{-1} (W_Q^T B W_p + W_Q^T J_p f W_p)$$

are invertible. This is always the case in the experiments that we considered.

5   Numerical Results

All numerical results are based on three simple network models which will be discussed in detail first.

Models

Pipeline

The first model is just one pipe of length 1000 m with a diameter of 1 m. Furthermore, we assume that the constants needed to compute the simulation are given by Table 1.

Table 1 Friction factor λ and isothermal speed of sound c_k used for the simulations

	Pipeline	Forked	Network
λ	0.0326	0.0326	0.05
c_k (in m²/s²)	300	300	430.5

Fig. 2 This is the topology of the test network

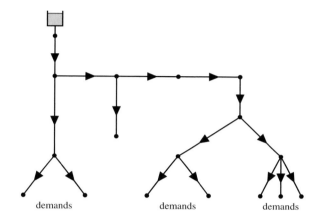

demands   demands   demands

We will always start the simulation from a steady state, which is given by a constant pressure of 70 bar at the inlet (left end of the pipe) and a constant mass flux at the outlet (right end of the pipe) of 60 kg/s.

Forked Network

The most simple pipe network that is not just a pipe is given by a fork, consisting of three pipes. On the left, there is an inlet node which then forks into two pipes and two outlet nodes. In total, we therefore have four nodes: one inlet given as a supply node, two outlet nodes, and one interior node at the fork. We have three pipes, which all have a length of 300 m and diameter 1 m. The other parameters are again given in Table 1. We assume that we always start our simulation in the steady state given by a supply pressure of 70 bar and both demand fluxes are 30 kg/s.

Network with 16 pipes

The topology of the last network is given in Fig. 2. We again assume that all pipes have the same length of 1000 m and diameter 1 m, and the parameters are given in Table 1. This network has one supply and eight demand nodes and eight interior nodes. We start our simulation in the steady state given by a supply pressure of 100 bar and all the demand fluxes at 60 kg/s.

Table 2 Steady-state solution for the flux at the inlet and the pressure at the outlet of the pipeline with 70 bar constant inlet pressure and 60 kg/s constant outlet flux for different space discretization sizes and the relative error compared to the finest discretization

	Flux	Pressure	Flux error	Pressure error
h = 250 m	58.999999999998252	69.990524501546957	0.29 1e-12	0.86 1e-5
h = 100 m	58.999999999996987	69.990888151088257	0.31 1e-12	0.34 1e-5
h = 50 m	59.000000000001634	69.991009363690068	0.23 1e-12	0.17 1e-5
h = 25 m	59.000000000013934	69.991069969257452	0.02 1e-12	0.08 1e-5
h = 25 m	58.999999999981789	69.991106332363344	0.57 1e-12	0.03 1e-5
h = 5 m	59.000000000028990	69.991118453359434	0.23 1e-12	0.01 1se-5

Input and Output

In all numerical experiments, we will talk about the input and the output function. The input functions are the pressure at the supply nodes and the mass flux at the demand nodes, and the output functions are the mass flux at the supply nodes and the pressure at the demand nodes.

DAE integration

Whenever the index of the DAE we are interested in simulating is 1 or lower, we use the MATLAB® standard ODE solver ode15s and otherwise we use a simple self-written solver based on an implicit–explicit splitting as in [10].

Discretization Effects

In this first numerical test, we show that refining the discretization seems to converge, and that a coarse discretization, which is often used in the literature, does give good results. We take the pipeline, and for a given discretization size we compute the steady-state solution. We compare the error at the output function. We assume that we start in a steady state and that the input is constant, which means that we remain in that steady state. In Table 2, the steady-state solution for the output, which is the mass flux at the inlet and the pressure at the outlet, is given for different discretization sizes. We furthermore compute the error between the value of the finest discretization and the given discretization, where in the finest discretization the discretization size is $h = 1$ m.

We see that even for the coarsest discretization we already get an accurate value for the flux at the inlet, which we are not able to improve by refining the discretization. For the value of the pressure at the outlet, we can see an improvement with increasing discretization; however, the solution of the coarsest discretization seems to be good enough.

Comparison of the Original and Approximated DAE

In this paper, we showed that the approximation introduced in [13] leads to a reduction of the DAE index. The advantage of this lower index is that many standard ODE solvers can still be used, whereas the integration of higher order index DAEs requires specialized DAE solvers. We compute the maximal relative error over time and all given outputs for all three models and different scenarios, meaning different input functions to show that the approximation does not introduce a significant error. We describe the scenarios, which means we describe the input function we use to run the transient simulations.

Pipe Scenarios

As mentioned before, we start in the steady state given above which for the pipe means pressure 70 bar on the left end and flux 60 kg/s on the right end. We simulate three scenarios. In the first scenario, we vary the flux by a sinusoidal function $q_s(t) = 60\,\text{kg/s} + sin(t/10)\,\text{kg/s}$. In the second scenario, we lower the flux linearly to 59 kg/s over 100 s and leave it there, and in the third scenario, we increase the pressure to 72 bar over 100 s.

Forked Scenarios

Again we start in the steady state mentioned above. In Scenario 1, we vary the flux of one demand node by $q_s(t) = 30\,\text{kg/s} + sin(t/10)\,\text{kg/s}$, and in Scenario 2 we lower the flux of one demand node linearly to 29 over 100 s and leave it there, and in Scenario 3 we increase the pressure to 72 bar over 100 s.

Network Scenarios

For the network, we only simulate two scenarios, where in Scenario 1 we vary the flux of all demands by $q_s(t) = 60\,\text{kg/s} + sin(t/10)$ and in Scenario 2 we increase the pressure of the supply node by 2 bar to 102 bar linearly over 100 s.

Results

We simulate all these scenarios for 200 s and compute the error, maximal relative error over time in all output functions. The results are summarized in Table 3.

In general, we see that the computation time for the original DAE is typically higher, where for the more complicated scenarios the speedup factor is about 20. The time integration scheme implemented for the index 2 DAE is certainly not optimal, but it is doubtful that one could achieve a speedup of 20 with a better implementation.

Model Order Reduction

Pipeline

In a first simple proof of concept, we use the pipeline model. We use a discretization size of $h = 10$ m, which leads to an overall system size of 200. The top two panels of Fig. 3 show the two input functions for the here computed scenario, as difference to the steady state input. The pressure at the supply node is constant, and the demand flux varies as a sinusoidal function. In the one below, we can see the two output functions for this scenario computed with the direct DAE model and with the approximation, the mass flow at the supply node and the pressure at the demand node. The absolute error is then plotted underneath.

The solution of this simulation is used to create the POD projection basis. We used a structured as well as an unstructured projection to create the reduced models. This means we get four reduced order models, two derived from each original DAE model, a structured and an unstructured one. We then simulated a different scenario. The input functions are again plotted on the top panel of Fig. 4. The sizes of the reduced

Table 3 Simulation time of a 200 s forward simulation for the directly discretized model and the approximation based on [13]. The speedup of the approximation and the maximal relative error between the two simulation results over time and all output functions

	Simulation time in s based on (5)	Simulation time in s based on (7)	Speedup	Error
Scenario 1 pipe	0.7	1	0.7	$1e-3$
Scenario 2 pipe	7.7	1.3	6	$1e-5$
Scenario 3 pipe	3	0.7	4	$1e-2$
Scenario 1 forked	10	1	10	$1e-4$
Scenario 2 forked	106	2.4	44	$1e-5$
Scenario 3 forked	47	2.1	22	$1e-3$
Scenario 1 network	12	5	2	$1e-3$
Scenario 2 network	526	28 s	18	$1e-2$

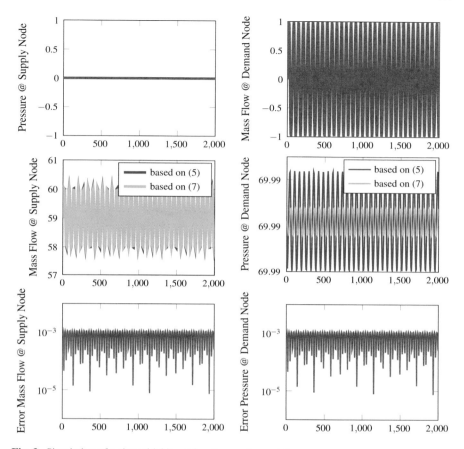

Fig. 3 Simulation of a sinusoidal input function as demand flux

model are picked such that whenever we do perform a singular value decomposition we truncate the singular values that are smaller than 10^{-4}. Therefore, the sizes of the reduced order models vary here. Also, the reduced order model created from the approximation with an unstructured projection did not lead to a meaningful simulation result and is therefore not included in the plot. In the second row of Fig. 4, we see the simulation results of the output for the different model and in the third row again the absolute error between the full and the reduced model, as well as the error between the direct and approximated model for comparison.

Network

Furthermore, we use the network consisting of 16 pipes. We use a discretization size of h = 50 m, and because each pipe is of length 1000 m the resulting system is of

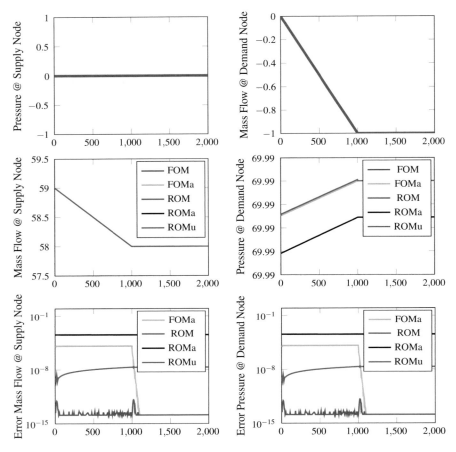

Fig. 4 Input and output functions given by five models: the full order model based on (5) (FOM) and the full order model based on the approximation (7) (FOMa), the structured reduced order models (ROM, ROMa), respectively, and the unstructured reduced model (FOMu) based on (5) and error plots comparing the reduced order model with its full order model

size 640. In this numerical experiment, we take the same scenario to create the POD basis and to simulate the reduced model to compare the error. The scenario we use is a step function in the pressure such that the pressure increases by 1 bar at 500 s and again by 1 bar at 1000 s and then goes back down 1 bar to 101 bar at 1500 s. We simulate for 2000 s. In Fig. 5, the mean relative output error is plotted over the size of the reduced order as well as the speedup.

Fig. 5 Mean relative output error of full and reduced model and computational speedup

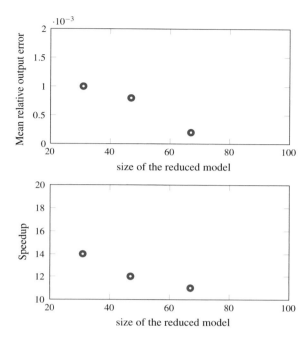

6 Conclusions

In this paper, we showed that the description given by (7) which is based on an approximation given in [13] results in a DAE that has more algebraic equations but is of index less than that of the original description. We can see that this also results in a faster simulation time and due to the lower index can also be reduced with standard model reduction techniques, which results in an even faster simulation.

Acknowledgements Financial support by the German Ministry of Economics (BMWi) within the Project MathEnergy, FKZ 03240198, within the research network energy system analysis (Energiesystemanalyse), is gratefully acknowledged. Responsibility for the contents of this paper rests with the authors.

References

1. Ali, G., Banagaaya, N., Schilders, W.H.A., Tischendorf, C.: Index-aware model order reduction for linear index-2 daes with constant coefficients. SIAM J. Sci. Comput. **35**(3), A1487–A1510 (2013)
2. Antoulas, A.C.: Approximation of Large-Scale Dynamical Systems. Advances in Design and Control, vol. 6. SIAM Publications, Philadelphia, PA (2005)

3. Baumann, M., Benner, P., Heiland, J.: Space-time galerkin pod with application in optimal control of semi-linear parabolic partial differential equations Cornell University (2016). math.OC. arXiv e-prints arXiv:1611.04050
4. Benner, P., Mehrmann, V., Sorensen, D.C.: Dimension Reduction of Large-Scale Systems. Lecture Notes in Computational Science and Engineering, vol. 45. Springer, Berlin (2005)
5. Domschke, P., Gross, M., Hante, F.M., Hiller, B., Schewe, L., Schmidt, M.: Mathematische modellierung, simulation und optimierung von gasnetzwerken. gwf Gas + Energie **11** (2015)
6. Ebert, F.: A note on POD model reduction methods for DAEs. Math. Comput. Model. Dyn. Syst. **16**(2), 115–131 (2010)
7. Ehrhardt, K., Steinbach, M.C.: Nonlinear optimization in gas networks. ZIB-Report ZR-03-46 (2003)
8. Fügenschuh, A., Geißler, B., Gollmer, R., Morsi, A., Pfetsch, M.E., Rövekamp, J., Schmidt, M., Spreckelsen, K., Steinbach, M.C.: Physical and technical fundamentals of gas networks. In: Koch, T., Hiller, B., Pfetsch, M.E., Schewe, L. (eds.) Evaluating Gas Network Capacities. SIAM Series on Optimization (2015)
9. Grundel, S., Hornung, N., Klaassen, B., Benner, P., Clees, T.: Computing surrogates for gas network simulation using model order reduction. In: Koziel, S., Leifsson, L. (eds.) Surrogate-Based Modeling and Optimization, pp. 189–212. Springer, New York (2013)
10. Grundel, S., Jansen, L.: Efficient simulation of transient gas networks using IMEX integration schemes and MOR methods. In: 54th IEEE Conference on Decision and Control (CDC), Osaka, Japan, pp. 4579–4584 (2015)
11. Grundel, S., Jansen, L., Hornung, N., Clees, T., Tischendorf, C., Benner, P.: Model order reduction of differential algebraic equations arising from the simulation of gas transport networks. In: Progress in Differential-Algebraic Equations. Differential-Algebraic Equations Forum, pp. 183–205. Springer, Berlin (2014)
12. Haasdonk, B.: Convergence rates of the POD-Greedy method. ESAIM. Math. Model. Numer. Anal. **47**(3), 859–873 (2013)
13. Herty, M., Mohring, J., Sachers, V.: A new model for gas flow in pipe networks. Math. Methods Appl. Sci. **33**(7), 845–855 (2010)
14. Jansen, L., Tischendorf, C.: A unified (p)dae modeling approach for flow networks. In: Schöps, S., et al. (ed.) Progress in Differential-Algebraic Equations, chap. 7, pp. 127–151. Springer (2014)
15. Kerschen, G., Golinval, J.-C., Vakakis, A.F., Bergman, L.A.: The method of proper orthogonal decomposition for dynamical characterization and order reduction of mechanical systems: an overview. Nonlinear Dyn. **41**, 147–169 (2005)
16. Kunisch, K., Volkwein, S., Xie, L.: HJB-POD-based feedback design for the optimal control of evolution problems. SIAM J. Appl. Dyn. Syst. **3**(4), 701–722 (2004)
17. Lamour, R., März, R., Tischendorf, C.: Differential-Algebraic Equations: A Projector Based Analysis. Springer Science & Business Media (2013)
18. März, R., Riaza, R.: Linear differential-algebraic equations with properly stated leading term: a-critical points. Math. Comput. Model. Dyn. Syst. **13**(3), 291–314 (2007)
19. Mehrmann, V.: Index concepts for differential-algebraic equations. In: Encyclopedia of Applied and Computational Mathematics, pp. 676–681. Springer (2015)
20. Pinnau, R.: Model reduction via proper orthogonal decomposition. In: Model Order Reduction: Theory, Research Aspects and Applications, pp. 95–109. Springer (2008)
21. Reddy, J.N.: An Introduction to Continuum Mechanics. Cambridge University Press, 2nd edn (2013)
22. Rowley, C.W., Colonius, T., Murray, R.M.: Model reduction for compressible flows using POD and Galerkin projection. Phys. D: Nonlinear Phenom. **189**(1–2), 115–129 (2004)
23. Rozza, G., Huynh, D.B.P., Patera, A.T.: Reduced basis approximation and a posteriori error estimation for affinely parametrized elliptic coercive partial differential equations. Arch. Comput. Methods Eng. **15**(3), 229–275 (2008)
24. Schilders, W.H.A., van der Vorst, H.A., Rommes, J.: Model Order Reduction: Theory, Research Aspects and Applications. Springer, Berlin (2008)

25. Schmidt, M., Steinbach, M.C., Willert, B.M.: High detail stationary optimization models for gas networks—part 1: model components. IfAM preprint **94** (2012)
26. Tröltzsch, F., Volkwein, S.: POD a-posteriori error estimates for linear-quadratic optimal control problems. Comput. Optim. Appl. **44**(1), 83–115 (2009)
27. Wyatt, S.: Issues in interpolatory model reduction: inexact solves, second order systems and DAEs. Ph.D. thesis, Virginia Polytechnic Institute and State University, Blacksburg, Virginia, USA (2012)

Model Order Reduction for Rotating Electrical Machines

Zeger Bontinck, Oliver Lass, Oliver Rain and Sebastian Schöps

Abstract The simulation of electric rotating machines is both computationally expensive and memory intensive. To overcome these costs, model order reduction techniques can be applied. The focus of this contribution is especially on machines that contain non-symmetric components. These are usually introduced during the mass production process and are modeled by small perturbations in the geometry (e.g., eccentricity) or the material parameters. While model order reduction for symmetric machines is clear and does not need special treatment, the non-symmetric setting adds additional challenges. An adaptive strategy based on proper orthogonal decomposition is developed to overcome these difficulties. Equipped with an a posteriori error estimator, the obtained solution is certified. Numerical examples are presented to demonstrate the effectiveness of the proposed method.

1 Introduction

Model order reduction for partial differential equations is a very active field in applied mathematics. When performing simulations in 2D or 3D using the finite element method (FEM), one arrives at large-scale systems that have to be solved. Projection-based model order reduction methods have shown to significantly reduce the

Z. Bontinck (✉) · S. Schöps
Technische Universität Darmstadt, Graduate School of Computational Engineering,
Dolivostr. 15, 64293 Darmstadt, Germany
e-mail: bontinck@gsc.tu-darmstadt.de

S. Schöps
e-mail: schoeps@gsc.tu-darmstadt.de

O. Lass
Department of Mathematics, Chair of Nonlinear Optimization,
Technische Universität Darmstadt, Dolivostr. 15, 64293 Darmstadt, Germany
e-mail: lass@mathematik.tu-darmstadt.de

O. Rain
Robert Bosch GmbH, 70049 Stuttgart, Germany
e-mail: oliver.rain@de.bosch.com

© Springer International Publishing AG, part of Springer Nature 2018
W. Keiper et al. (eds.), *Reduced-Order Modeling (ROM) for Simulation and Optimization*,
https://doi.org/10.1007/978-3-319-75319-5_6

computational complexity when applied carefully. While being used in many different fields in physics, the application to rotating electrical machines is more recent [10, 14, 18, 30]. We will focus especially on the setting of non-symmetric machines. While the perfect machine is symmetric and simulations are usually carried out exploiting these properties, in real life the symmetry is often lost. This is due to perturbation in the geometry (e.g., eccentricity) and material properties which requires that the whole machine, not only a small portion, is simulated (e.g., one pole). Hence this leads to an increase in the computational cost. The aim is to develop an adaptive strategy that is able to collect the required information systematically. Ideally, the algorithm is able to detect symmetries and exploits them if present. The greedy algorithm introduced in the context of the reduced basis method is a possible candidate [22, 26]. A commonly used method in engineering and applied mathematics is the method of snapshots or proper orthogonal decomposition (POD) [3, 11, 31]. We opt for a combination of these two methodologies. The goal is a fast and efficient algorithm that avoids the necessity of an online–offline decomposition. This allows a broader application since no expensive offline costs have to be invested. Hence the method will already pay off after one simulation and not only in the many query context.

Additionally, the developed strategy has to be able to handle the motion of the rotor. While there are a number of methods to treat the rotation [6, 23, 29], we will assume a constant rotational speed which allows us to utilize the locked step method [24]. Hence we can avoid the remeshing which is computationally prohibitive. Moreover, the application of other approaches should be straight forward.

Efficient simulation tools are a key ingredient when performing optimization or uncertainty quantification. The combination of model order reduction and optimization has caught a lot of attention; see for example [7, 8, 19, 25, 33]. Especially in the many query context, where models have to be evaluated repeatedly, there is a need for an accurate, fast and reliable reduced order model. While we will not look into the application of the reduced order models, we will develop a strategy that fulfills these needs. By using existing simulation tools in the adaptive procedure, it is possible to insert the developed strategy into an existing framework and utilize the benefits of the reduced order model, as shown in [14].

The article is structured as follows: In Sect. 2 the permanent magnet synchronous machine (PMSM) is introduced and discretized. The model order reduction strategy is developed in Sect. 3. Then in Sect. 4, numerical experiments are presented. Lastly, a conclusion is drawn in the last section.

2   Problem Description

This section is devoted to the modeling of the PMSM. Furthermore, we introduce the finite element discretization and outline the realization of the rotation for the discrete setting.

Model Problem

The PMSM under investigation has six slots per pole and a double-layered winding scheme with two slots per pole per phase. The geometry of the full machine is shown in Fig. 1 (left) together with a detailed view on one pole (right). In each pole there is one buried permanent magnet indicated in gray. The machine has depth $\ell_z = 10$ mm. It is operated at 50 Hz, resulting in a rotational speed of 1000 revolutions per minute (RPM). The machine is composed of laminated steel with a relative permeability $\mu_r = 500$. In the following, Ω_s and Ω_r refer to the stator and rotor domains, respectively. The whole domain is then referred to as $\Omega = \Omega_s \cup \Omega_r$ and will be used for simpler notation when appropriate. Additionally, let us define the interface $\Gamma_I = \partial \Omega_s \cap \partial \Omega_r$ (dashed line) in the airgap between the rotor and the stator. Furthermore, we introduce the boundaries $\Gamma_s = \partial \Omega_s \setminus \Gamma_I$ and $\Gamma_r = \partial \Omega_r \setminus \Gamma_I$ of the stator and rotor, correspondingly. In the simulation we will account for the movement of the rotor, hence we introduce the angle ϑ that describes the position of the rotor with respect to the stator. For clarity we will append ϑ to the components related to the rotor.

To calculate the magnetic vector potential of the machine, the magnetostatic approximation of the Maxwell's equations has to be solved for both domains. This implies that the eddy and displacement currents are neglected and one obtains the semi-elliptic partial differential equations

$$\nabla \times (\nu \nabla \times \mathbf{A}(\vartheta)) = \mathbf{J}_{src}(\vartheta) - \nabla \times (\nu \mathbf{B}_{rem}), \quad \text{on } \Omega \tag{1}$$

with Dirichlet boundary conditions $\mathbf{A} \times \mathbf{n} = 0$ on Γ_s, with $\mathbf{n}$ the outer unit normal. The reluctivity is depicted by a scalar ν since only linear isotropic materials are considered and nonlinearity is disregarded since a linearization at a working point is assumed. $\mathbf{A}(\vartheta)$ is the magnetic vector potential, $\mathbf{J}_{src}(\vartheta)$ represents the imposed source current density, which is related to the applied currents in the coils, and $\mathbf{B}_{rem}$ is the remanence of the permanent magnets. The applied current density is aligned with

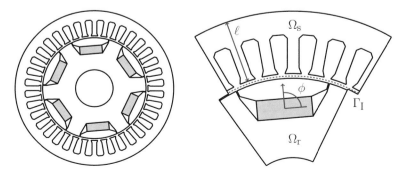

Fig. 1 Cross-sectional view of the full PMSM (left) and detailed view on one pole (right)

the z-direction, whereas the remanence is in the xy-plane. It is generally accepted that machines are adequately modeled in 2D, meaning it is assumed that the magnetic vector potential only has a z-component: $\mathbf{A} = (0, 0, A_z)$. Since $\mathbf{B} = \nabla \times \mathbf{A}$, one can write for the magnetic field: $\mathbf{B} = (B_x, B_y, 0)$. Hence we end up with the linear elliptic equation

$$- \nabla \cdot (\nu \nabla A_z(\vartheta)) = j_{\mathrm{src}}(\vartheta) - j_{\mathrm{pm}}, \quad \text{on } \Omega(\vartheta), \tag{2}$$

where j_{src} and j_{pm} are the z-component of $\mathbf{J}_{\mathrm{src}}$ and $\nabla \times (\nu \mathbf{B}_{\mathrm{rem}})$, respectively. Further, the boundary conditions are given as previously introduced by $A_z = 0$ on Γ_{s}. For simplicity we collect all the terms on the right side in a function f. The weak form of (2) can be written as

$$\int_{\Omega(\vartheta)} \nu \nabla A_z(\vartheta) \nabla w \mathrm{d}\mathbf{x} = \int_{\Omega(\vartheta)} f(\vartheta) w \mathrm{d}\mathbf{x}, \quad \forall w \in H_0^1(\Omega(\vartheta)). \tag{3}$$

where $w \in H_0^1(\Omega(\vartheta))$ are the test functions. This formulation will be the starting point for the finite element discretization in the next section.

Let us next have a look at possible imperfections in the presented geometry and model. These are introduced during mass production of the PMSMs and are given by small deviations in the geometry or material properties. While there are many possible imperfections that can occur, we focus on two types. On the one hand we look at imperfection in the material of the permanent magnet, more precisely we consider deviations in the magnetic field angle ϕ of the permanent magnet [20]. The second imperfection we consider is the length ℓ of the teeth in the stator [5, 21]. Both quantities are depicted in Fig. 1 (right). For the shown PMSM, we have 36 teeth in the stator and 6 permanent magnets. Each tooth length and magnetic field angle can be perturbed independently, hence this results in a system depending on 42 parameters, i.e., $l \in \mathbb{R}^{36}$ and $\phi \in \mathbb{R}^6$. Perturbations in these quantities may lead to under performance of the PMSM.

Finite Element Discretization

We obtain the discrete version of (3) by utilizing the finite element method (FEM). By discretizing A_z by linear edge shape functions $w_i(x, y)$, one makes the Ansatz

$$A_z^{\mathrm{N}} = \sum_{i=1}^{N} a_i w_i(x, y) = \sum_{i=1}^{N} a_i \frac{\varphi_i(x, y)}{\ell_z} e_z,$$

where $\varphi_i(x, y)$ depicts the nodal finite elements which are associated with the triangulation of the machine's cross section and e_z is the unit vector in z-direction. Inserting this Ansatz in (3) and using w_i as test functions in the Ritz–Galerkin approach, the resulting N-dimensional linear discrete system is given by

$$\mathbf{K}_\nu(\vartheta)\mathbf{a}(\vartheta) = \mathbf{f}(\vartheta), \tag{4}$$

where $\mathbf{K}_\nu$ are the finite element system matrices, $\mathbf{a}$ depict the degrees of freedom (DoFs), and $\mathbf{f} = \mathbf{j}_{\mathrm{src}} - \mathbf{j}_{\mathrm{pm}}$ are the discretized versions of the current densities and permanent magnets.

To take the motion of the rotor into account in the simulation, we utilize the locked step method [24, 29]. For the implementation, a contour in the airgap is defined (Γ_I) which splits the full domain into two parts: a fixed outer domain connected to the stator Ω_s and an inner domain connected to the rotor Ω_r, where the mesh will be rotated. At the contour, the nodes are distributed equidistantly. The angular increments $\Delta\vartheta$ are chosen so that the mesh of the stator and rotor will match on the interface. The nodes on the interface are then reconnected leading to the mesh for the next computation. Using this technique, the rotation angle ϑ is given by $\vartheta^k = k\Delta\vartheta$ with $k \in \mathbb{N}_0$. We can hence partition the discrete unknown $\mathbf{a}$ into a static part, a rotating part, and the interface, with dimensions N_s, N_r, and N_I, respectively. This idea is a particularization of non-overlapping domain decomposition [1, 32]. The linear system (4) can then be written as

$$\begin{bmatrix} \mathbf{K}_\nu^{ss} & 0 & \mathbf{K}_\nu^{sI} \\ 0 & \mathbf{K}_\nu^{rr} & \mathbf{K}_\nu^{rI}(\vartheta) \\ (\mathbf{K}_\nu^{sI})^\top & (\mathbf{K}_\nu^{rI})(\vartheta)^\top & \mathbf{K}_\nu^{II}(\vartheta) \end{bmatrix} \begin{bmatrix} \mathbf{a}_s \\ \mathbf{a}_r \\ \mathbf{a}_I \end{bmatrix} = \begin{bmatrix} \mathbf{f}^s \\ \mathbf{f}^r \\ \mathbf{f}^I(\vartheta) \end{bmatrix}, \tag{5}$$

where $\mathbf{K}_\nu^{ss}$, $\mathbf{K}_\nu^{rr}$, $\mathbf{f}^s$, and $\mathbf{f}^r$ are the stiffness matrices and right-hand sides on the static and moving part, which no longer depend on the angle ϑ. For the points on the interface, there are two cases. The interface of the static part is independent of the angle ϑ and hence we get the corresponding stiffness matrix $\mathbf{K}_\nu^{sI}$. For the rotor side we have to perform the shift; this is indicated with ϑ in the corresponding stiffness matrix $\mathbf{K}_\nu^{rI}$. On the interface, also a shift has to be performed; hence also here the corresponding stiffness matrix $\mathbf{K}_\nu^{II}$ and right-hand side $\mathbf{f}^I$ are dependent on ϑ. Let us note that it is not required to reassemble matrices. All of these shifts can be performed by index shift and hence allow a very efficient implementation. Moreover, the size of the system does not change, i.e., we have $\mathrm{N} = \mathrm{N}_s + \mathrm{N}_r + \mathrm{N}_I$.

3   Model Order Reduction

The linear system (5) resulting from the finite element discretization is of large scale and hence expensive to solve. To reduce the computational costs, model order reduction will be utilized. In this section, we present a model order reduction framework based on proper orthogonal decomposition to speedup the simulations.

Reduced Order Model

The goal is to generate a reduced order model to accelerate the simulation of (5). The simulation of the rotation is computationally expensive since the discretization of (2) leads to very large linear systems that need to be solved for every angular position. While in symmetric machines this can be avoided, in the case of non-symmetric machines a whole revolution has to be simulated. Hence we require an efficient strategy to overcome this problem. For this we investigate an adaptive strategy that builds a surrogate model while performing the simulation and switches to it when the required accuracy is reached. We want to use information collected over the rotational angle ϑ and generate a projection-based reduced order model. In the past, model order reduction methods based on proper orthogonal decomposition (POD) [3, 11, 13], balanced truncation [1, 12], and the reduced basis method [22, 26, 27] have been developed to speedup the computation. More recently, the POD method has been successfully applied to rotating machines [10, 14, 18, 30]. In this work, we consider a combination of POD and the reduced basis method.

We will not pursue an online–offline decomposition but rather see the reduced order model as an accelerator for the simulation. Further, we will only generate the reduced order model with respect to the rotation. The reason for this is the large number of parameters that can occur in the model. In the presented setting, we deal with 36 geometry parameters, 6 material parameters, and the rotation. With this large parameter space, a classical reduced basis strategy with online–offline decomposition becomes infeasible. Furthermore, the goal is to employ these reduced order models in a parametrized shape optimization framework. The goal is to speed up the simulation in each optimization iteration when the optimization variables and geometry/material parameters are fixed [14].

If it is desired to generate a reduced order model involving the parameters ℓ, ϕ, and ϑ, this can be achieved by a classical reduced basis approach. There exist efficient methods depending on how the parameters enter the equation for low-dimensional parameter spaces. Especially, in the case that the parametrization of the geometry can be decomposed using a linear affine decomposition, this can be an interesting option [22, 27]. If the geometry transformation is nonlinear, an online–offline decomposition is possible using the empirical interpolation method (EIM) [2, 4]. Note that in the investigated setting, we do not have any assumptions on how the parameters enter the equation and the parameter space is of high dimension. Hence we focus primarily on the reduction of the computational cost with respect to the rotation.

Proper Orthogonal Decomposition

Let us start by recalling the POD method so we can develop an extension suitable for the application presented. Let the solution to (2) be given in the discrete form, i.e., let $\mathbf{a}(\vartheta) \in \mathbb{R}^N$ be the solution to (4) for a fixed angle ϑ. The snapshots are then given by $\mathbb{R}^N \ni \mathbf{a}^k \approx \mathbf{a}(k\Delta\vartheta)$ for $k \in \mathcal{K}$, where $\mathcal{K}$ is an index set with elements in $\mathbb{N}_0$ for which (4) is solved. A POD basis $\{\psi_i\}_{i=1}^{\ell}$ is then computed from these snapshots by solving the following optimization problem:

$$\begin{cases} \min_{\psi_1,\ldots,\psi_{\mathcal{N}} \in \mathbb{R}^N} \sum_{k \in \mathcal{K}} \left| \mathbf{a}^k - \sum_{i=1}^{\mathcal{N}} \langle \mathbf{a}^k, \psi_i \rangle_{\mathrm{W}} \, \psi_i \right|_{\mathrm{W}}^2 \\ \text{s.t. } \langle \psi_i, \psi_j \rangle_{\mathrm{W}} = \delta_{ij} \text{ for } 1 \leq i, j \leq \mathcal{N}, \end{cases} \tag{6}$$

where $\langle \cdot, \cdot \rangle_{\mathrm{W}}$ stands for the weighted inner product in $\mathbb{R}^N$ with a positive definite, symmetric matrix $\mathrm{W} \in \mathbb{R}^{N \times N}$. The goal is to minimize the mean projection error of the given snapshots projected onto the subspace spanned by the POD basis ψ_i. By introducing the matrix $\mathbf{A}_{\mathcal{K}}$ as the collection of the snapshots $\mathbf{a}^k$ with $k \in \mathcal{K}$, we can write the operator $\mathbf{R}$ arising from the optimization problem (6) as

$$\mathbf{R}\psi = \sum_{k \in \mathcal{K}} \langle \mathbf{a}^k, \psi \rangle_{\mathrm{W}} \mathbf{a}^k = \mathbf{A}_{\mathcal{K}} \left(\mathbf{A}_{\mathcal{K}} \right)^{\top} \mathrm{W}\psi \quad \text{for } \psi \in \mathbb{R}^N.$$

Then the unique solution to (6) is given by the eigenvectors corresponding to the $\mathcal{N}$ largest eigenvalues of $\mathbf{R}$, i.e., $\mathbf{R}\psi_i = \lambda_i \psi_i$ with $\lambda_i > 0$ [8]. The operator $\mathbf{R}$ is large since it is of dimension N which we want to reduce. Hence it might be better in many cases to set up and solve the eigenvalue problem

$$\mathbf{A}_{\mathcal{K}}^{\top} \mathrm{W} \mathbf{A}_{\mathcal{K}} v_i = \lambda_i v_i, \quad i = 1, \ldots, \mathcal{N}$$

and obtain the POD basis by $\psi_i = 1/\sqrt{\lambda_i} \mathbf{A}_{\mathcal{K}} v_i$. Note that both approaches are equivalent and are related by the singular value decomposition (SVD) of the matrix $\mathrm{W}^{1/2}\mathbf{A}_{\mathcal{K}}$. While the eigenvalue decomposition is computationally more efficient, the SVD is numerically more stable. A comparison of the different computations was carried out in [15]. For completeness, let us state the POD approximation error given by

$$\sum_{k \in \mathcal{K}} \left| \mathbf{a}^k - \sum_{i=1}^{\mathcal{N}} \langle \mathbf{a}^k, \psi_i \rangle_{\mathrm{W}} \, \psi_i \right|_{\mathrm{W}}^2 = \sum_{i=\mathcal{N}+1}^{d} \lambda_i, \tag{7}$$

where d is the rank of $\mathbf{A}_{\mathcal{K}}$. For easier notation, we collect the POD basis ψ_i in the matrix $\Psi = [\psi_1, \ldots, \psi_{\mathcal{N}}] \in \mathbb{R}^{N \times \mathcal{N}}$.

After introducing the computation of the POD basis, we will now outline the adaptive approach utilized in this work. It is crucial to minimize the number of solves involving the FEM discretization in order to obtain a speedup of the computation. The goal is to push most of the computations in the simulation onto the reduced order models. However, we have to guarantee that the reduced order models are accurate in order to obtain reliable results. We will present a strategy that does not require precomputation as for example in the reduced basis method but performs the model order reduction during the simulation. First let us outline how the POD basis is applied to (5). In the second step we give the details on how to obtain the basis efficiently.

We generate for each part of the machine an individual POD basis. Hence we have one basis for the stator and one basis for the rotor. The interface between the stator and the rotor is not reduced, i.e., we work with the FEM Ansatz space on the interface. This is motivated by the observation that the decay of the eigenvalues on Γ_{I} is very slow, which would result in a large POD basis. Since the FEM space for the interface is usually of moderate dimension, the gain of using POD would be negligible. We compute the POD basis as a solution to (6) utilizing the snapshots $\mathbf{a}_s^k$ and $\mathbf{a}_r^k$ to obtain Ψ^s and Ψ^r, respectively. We then make the Ansatz

$$\mathbf{a}_s^{\mathcal{N}} = \sum_{i=1}^{\mathcal{N}_s} \psi_i^s \bar{\mathbf{a}}_{s,i} = \Psi^s \bar{\mathbf{a}}_s \quad \text{and} \quad \mathbf{a}_r^{\mathcal{N}} = \sum_{i=1}^{\mathcal{N}_r} \psi_i^r \bar{\mathbf{a}}_{r,i} = \Psi^r \bar{\mathbf{a}}_r,$$

where the POD coefficients are indicated with a bar. By projecting (5) onto the subspace spanned by the POD basis, we obtain the reduced order model

$$\begin{bmatrix} (\Psi^s)^\top \mathbf{K}_\nu^{ss} \Psi^s & 0 & (\Psi^s)^\top \mathbf{K}_\nu^{sI} \\ 0 & (\Psi^r)^\top \mathbf{K}_\nu^{rr} \Psi^r & (\Psi^r)^\top \mathbf{K}_\nu^{rI}(\vartheta) \\ (\mathbf{K}_\nu^{sI})^\top \Psi^s & (\mathbf{K}_\nu^{rI})^\top(\vartheta)\Psi^r & \mathbf{K}_\nu^{II}(\vartheta) \end{bmatrix} \begin{bmatrix} \bar{\mathbf{a}}_s \\ \bar{\mathbf{a}}_r \\ \mathbf{a}_I \end{bmatrix} = \begin{bmatrix} (\Psi^s)^\top \mathbf{f}^s \\ (\Psi^r)^\top \mathbf{f}^r \\ \mathbf{f}^I(\vartheta) \end{bmatrix}.$$

In short notation, the system will be written as

$$\mathbf{K}_\nu^{\mathcal{N}}(\vartheta)\bar{\mathbf{a}} = \mathbf{f}^{\mathcal{N}}(\vartheta) \tag{8}$$

with $\bar{\mathbf{a}}$ the vector of POD coefficients. This system is of dimension $\mathcal{N}_s + \mathcal{N}_r + \mathrm{N}_I$ and of much smaller dimension as the original system (5) which has N degrees of freedom.

Let us now shortly have a look at the error estimator. For this let us recall some basic quantities. We define the discrete coercivity constant by

$$\alpha(\vartheta) = \inf_{v \in \mathbb{R}^N \setminus \{0\}} \frac{\mathbf{v}^\top \mathbf{K}_\nu(\vartheta)\mathbf{v}}{\mathbf{v}^\top \mathbf{W} \mathbf{v}}.$$

Hence the coercivity constant is given by the smallest eigenvalue such that

$$\mathbf{K}_\nu(\vartheta)\mathbf{v} = \lambda \mathbf{W} \mathbf{v}$$

is satisfied for $(\lambda, \mathbf{v}) \in \mathbb{R}_+ \times \mathbb{R}^N$ and $\mathbf{v} \neq 0$ [26]. Further, we define the residual $r(\mathbf{a}^{\mathcal{N}}; \vartheta) = \mathbf{f}(\vartheta) - \mathbf{K}_\nu(\vartheta)\mathbf{a}^{\mathcal{N}}$. Then the error introduced by the reduced order model in the variable $\mathbf{a}$ can be characterized by

$$\|\mathbf{a} - \mathbf{a}^{\mathcal{N}}\|_{\mathbf{W}} \leq \Delta_{\mathbf{a}}(\vartheta) := \frac{\|r(\mathbf{a}^{\mathcal{N}}; \vartheta)\|_{\mathbf{W}^{-1}}}{\alpha(\vartheta)}. \tag{9}$$

Additionally, we look at the relative error which is more interesting in many applications. The corresponding error estimator reads

$$\frac{\|\mathbf{a} - \mathbf{a}^{\mathcal{N}}\|_{\mathbf{W}}}{\|\mathbf{a}^{\mathcal{N}}\|_{\mathbf{W}}} \leq \Delta_{\mathbf{a}}^{\mathrm{rel}}(\vartheta) := 2\frac{\|r(\mathbf{a}^{\mathcal{N}}; \vartheta)\|_{\mathbf{W}^{-1}}}{\alpha(\vartheta)\|\mathbf{a}^{\mathcal{N}}\|_{\mathbf{W}}}. \tag{10}$$

This is a standard result and can be found in [22, 27]. In the numerical realization we set the weight matrix $\mathbf{W}$ to the N dimensional identity matrix. Other choices (e.g., $\mathbf{W} = \mathbf{K}_\nu$) are possible but not investigated at this point. Note that the rotation as introduced in this work does not influence the coercivity constant and hence the dependence can be omitted. This can also be seen in the numerics, where only small deviations can be observed which are in the order of discretization. Hence a very efficient realization is possible, since only one eigenvalue problem has to be solved. Let us remark that the error is measured with respect to the finite element solution. It is assumed that the finite element solution is accurate enough to approximate the solution of the continuous problem.

Lastly let us have a look at how we choose the dimension $\mathcal{N}$ of the POD basis. We require that

$$\frac{\sum_{i=1}^{\mathcal{N}} \lambda_i}{\sum_{i=1}^{d} \lambda_i} \leq \varepsilon_{rel}$$

holds for the stator and rotor independently. This is a popular choice, where a typical value is $\varepsilon_{rel} = 0.9999$. Note that the denominator can be computed by $\mathrm{trace}(\mathbf{A}_{\mathcal{K}}^{\top}\mathbf{W}\mathbf{A}_{\mathcal{K}})$ and hence not all d eigenvalues have to be computed.

Adaptive Strategy

Next we introduce the strategy on how to determine the POD basis adaptively. The goal is to reduce the computational cost with respect to the rotation. A full revolution requires N_I solves of the system (5), i.e., for all ϑ_k with $k \in \mathcal{K} := \{0, 1, \ldots, N_I - 1\}$. In the symmetric case it is not required to solve a full rotation but only for angles that cover one pole, for our particular example this means one-sixth, i.e., $N_I/6$ solutions are needed. Note, we assume that N_I is divisible by N_p, where N_p is the number of poles of the machine. In the non-symmetric case this is not possible.

The idea is to partition the set $\mathcal{K}$ and generate a sequence of disjoint sets $\mathcal{K}_i$, $i = 0, \ldots, N_{\mathcal{K}}$, i.e.,

$$\mathcal{K}_i \cap \mathcal{K}_j = \emptyset, \quad i \neq j \quad \text{and} \quad \mathcal{K} = \bigcup_{i=1}^{N_{\mathcal{K}}} \mathcal{K}_i. \tag{11}$$

Here the sets can be chosen arbitrarily as long as they fulfill the introduced property. We require that the sets are disjoint to minimize computational overhead when generating the basis. Ideally the sets $\mathcal{K}_i$ are not too large but are large enough to capture the most important features. In the case of rotating machines, there are logical choices for these sets which we will outline in the numerical results.

The strategy is then as follows: We start by choosing $\mathcal{K}_0$ and evaluate (5) for ϑ_k and $k \in \mathcal{K}_0$. From the computed solutions/snapshots $A_{\mathcal{K}} = [\mathbf{a}^k]_{k \in \mathcal{K}_0}$, a POD basis is computed. Then the error estimator $\Delta_{\mathbf{a}}(\vartheta_k)$, $k = 1, \ldots, N_I$, is evaluated to determine the maximum error. If the error is larger than a given tolerance the index $k \in \{1, \ldots, N_I\}$ is determined, where the maximum error is attained. Next we determine the set $\mathcal{K}_i$ that contains the index k and the snapshot set is enlarged by adding the new solutions corresponding to $\mathcal{K}_i$ to the old ones, i.e., $A_{\mathcal{K}} = [A_{\mathcal{K}}, [\mathbf{a}^k]_{k \in \mathcal{K}_i}]$. This procedure is repeated until the error estimator $\Delta_{\mathbf{a}}(\vartheta_k)$, $k = 1, \ldots, N_I$, is below the desired threshold. In Algorithm 1, the strategy is summarized.

For stability reasons, the sets $\mathcal{K}_i$ that have been already used are removed from the list. In the case that the largest index is in a removed set, the index for the second largest error is used. Alternatively the number of basis functions can be increased by lowering the tolerance ε_{rel}, e.g., $\varepsilon_{rel} = \tau \varepsilon_{rel}$ with $\tau \in (0, 1)$. We have not encountered this scenario in our numerical experiments and hence it will not be investigated further.

This sampling of the sets $\mathcal{K}_i$ is similar to the greedy algorithm from the reduced basis method. The decision to add more than one solution $\mathbf{a}^k$ at a time to the snapshot set is to minimize the overhead of evaluating the error estimator and generating the reduced order models. Since we do not introduce an online–offline decomposition, the computations of error estimator and the generation of the reduced order models are included in the computational costs.

In the presented approach in each iteration of the strategy, an eigenvalue decomposition has to be computed. For large problems this can become computationally expensive. Alternatively, a hierarchical approach can be chosen. Then a POD basis is computed for each snapshot set independently and then merged by a Gram–Schmidt process. This is very similar to the approach used in the POD–Greedy setting introduced in [9].

Algorithm 1 Adaptive POD

Require: $\mathcal{K}_{i=0}^{N_{\mathcal{K}}}$ and ε (tolerance)
1: Choose first set, e.g., $i = 0$ and set the snapshot set $\mathbf{A}_{\mathcal{K}} = []$
2: Solve $\mathbf{a}^k$ for $k \in \mathcal{K}_i$ and add to $\mathbf{A}_{\mathcal{K}}$
3: Compute POD basis using (6)
4: Evaluate error estimator $\Delta_{\mathbf{a}}^{\text{rel}}(\vartheta_k)$ for $k = 1, \ldots, N_I$
5: **if** $\max_k \Delta_{\mathbf{a}}^{\text{rel}}(\vartheta_k) > \varepsilon$ is in same partition **then**
6: $\quad$ Determine index i of set $\mathcal{K}_i$ containing k
7: $\quad$ GOTO 2
8: **else**
9: $\quad$ **return** POD basis and reduced solution $\mathbf{a}^\ell$
10: **end if**

---

4 Numerical Results

We will now present different numerical results. For this let us specify the settings. The geometry (Fig. 1) is discretized using a triangular mesh with 56297 nodes, and the interface Γ_{I} is discretized by 900 equidistant points. Hence one revolution requires the linear system (4) to be solved 900 times. Here is where model order reduction will come into play and speedup the simulation significantly. All computations are performed on a standard desktop PC using MATLAB R2016b. Throughout our numerical tests, we will consider the following four settings:

sym $\quad$ Symmetric machine.
rot $\quad$ Perturbation of ϕ in one permanent magnet by $5°$.
$stat$ $\quad$ Perturbation of ℓ in one tooth by 0.3mm.
rot_stat $\quad$ Perturbation in both ϕ and ℓ.

In the numerical simulation, we only consider the perturbation of a single ϕ and a single ℓ, i.e., $\phi = [5, 0, \ldots, 0] \in \mathbb{R}^6$ and $\ell = [0.3, 0, \ldots, 0] \in \mathbb{R}^{36}$. This is to illustrate the influence of small perturbations in the design and material on the reduced order model. In the case that more than one value of ℓ and ϕ is perturbed, the performance of the algorithm is very similar. Due to the large number of parameters and combinations, we only investigate the presented settings which are very representative.

To start, we will have a look at the performance of the POD method. For this we compute a full revolution for each of the four settings and have a look at the decay of the eigenvalues. Let us recall that a fast decay is essential for the POD method to perform well. In Fig. 2 the decay of the normalized eigenvalues is shown. As can be seen, the eigenvalues decay very fast for rotor and stator. Only for the interface, the decay is much slower. This underlines the decision to not perform a model order reduction for the interface. To generate a good reduced order model, additional snapshots would be required which is computationally expensive. Since we do not have an online–offline decomposition, we have to minimize the number of computed snapshots. Additionally, it can be observed that a perturbation in one magnetic field angle ϕ has less influence on the decay of the eigenvalues than the

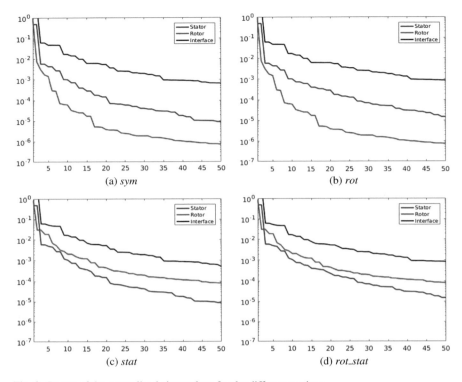

Fig. 2 Decay of the normalized eigenvalues for the different settings

perturbation in the length ℓ of one stator tooth. While the perturbation in ϕ causes a slight change in the decay of the eigenvalues related to the stator, the perturbation in ℓ dramatically influences the eigenvalues related to the rotor.

Next we depict the first three POD basis vectors for each setting (Fig. 3, 4, 5 and 6). For the setting *sym* and *rot*, very similar basis vectors are obtained. The last basis vector for both settings is zero on wide areas and only adds contributions at the interface (green is zero). In the settings *stat* and *rot_stat* this is very different. There the third basis vector for the rotor still contributes a lot of information to the reduced order model. What was already observed in the decay of the eigenvalues was verified again in the POD basis vectors. The settings *sym* and *rot* exhibit similar behaviors as well as *stat* and *rot_stat*.

From the decay of the eigenvalues, it can be expected that we will be able to generate reduced order models of very low dimension. The plots of POD basis verify this for the *sym* and *rot* since already the third basis vector is almost zero.

Next we have a look at the performance of the presented adaptive algorithm. For this let us introduce the required index sets. We will look at two settings. To be able to differentiate the two settings, we will call the sets $\mathcal{M}$ and $\mathcal{K}$. The presented sets do not require any prior knowledge of the machine other than the number of poles

Fig. 3 First three POD basis vectors for the stator (top) and rotor (bottom) for setting *sym*

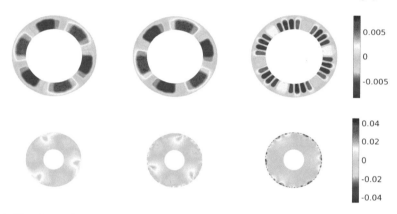

Fig. 4 First three POD basis vectors for the stator (top) and rotor (bottom) for setting *rot*

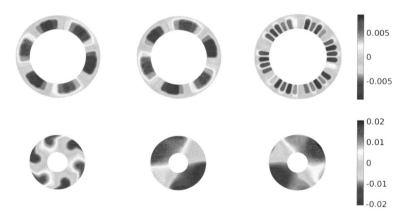

Fig. 5 First three POD basis vectors for the stator (top) and rotor (bottom) for setting *stat*

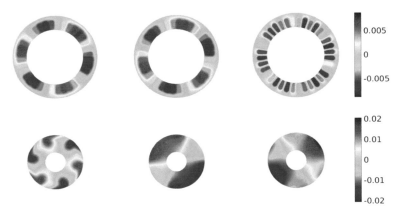

Fig. 6 First three POD basis vectors for the stator (top) and rotor (bottom) for setting *rot_stat*

and points on the interface and they are the logical choices. Note that also random sets or sets with only one element would be possible but we will not investigate this here.

The first strategy is to divide the full rotation into six sections (6 poles). This means we have 150 interface points in each section which are then partitioned into 12 partitions. The indices for each partition are equidistantly distributed over the section, hence we have a hierarchy of sets. The particular choice of sets can be written as follows

$$\mathcal{K}_{ij} = 150(i - 1) + [j : 12 : 150], \quad i = 1, \ldots, 6 \quad \text{and} \quad j = 1, \ldots, 12,$$

where $[j : k : n]$ is the notation for a regular spaced index set with increment k between j and n, i.e., $[j, j + k, j + 2k, \ldots, j + mk]$ with $m = \lfloor (n - j)/k \rfloor$, where $\lfloor \cdot \rfloor$ is the floor function. This leaves us with 12 sets for each pole and 72 sets for the whole machine. Using these sets, we always compute snapshots related only to the rotation angle covering one pole. When the algorithm chooses multiple sets associated to the same pole, this can be interpreted as a local refinement, i.e., more information for a specific section is added.

In the second strategy, distributed index sets are considered. This allows for a broader information collection since the snapshots are immediately distributed over the whole range of ϑ. To have a comparable setting with the $\mathcal{K}$ sets, we require again 72 partitions. We define the corresponding sets by

$$\mathcal{M}_i = [i : 72 : 900], \quad i = 1, \ldots, 72.$$

In this setting, the computed snapshots are not associated to local properties of the machine but are distributed. The indices in each set are distributed equidistantly over the whole rotation and unifying them can be interpreted as a global refinement.

In our experiment, it turned out that these settings are a good trade-off between performance and accuracy. Furthermore, these choices do not require any prior knowledge of the machine under investigation, but are choices of gathering snapshots in a local or global manner. As the tolerance for our adaptive strategy we use $\varepsilon = 10^{-3}$, which has been indicated as sufficient by the industry partner. The error estimator overestimated the real error by around one to two orders of magnitude. This is a good result since it guarantees that the model is not refined unnecessarily often.

The results for the two different index sets $\mathscr{K}$ and $\mathscr{M}$ are shown in Figs. 7 and 8. We plot the error estimator $\Delta_{\mathbf{a}}^{\mathrm{rel}}$ for the relative error of the 900 angular positions for every iteration of the adaptive algorithm. The actual error is omitted in the plots for visual clarity. By a dashed line, the desired tolerance is indicated as a visual aide. It can be seen how each of the two index sets has very different behaviors.

The set $\mathscr{K}$, which uses a kind of local refinement of the snapshots, performs well for the *sym* and *rot*. In the behavior of the error, it can be clearly seen in which region new snapshots were added. The error of the region drops significantly. As was expected, the symmetric case only requires one iteration since already a few snapshots contain enough information to compute the full rotation. For the perturbation in the rotor, a second set of snapshots is required to push the error below the tolerance ε.

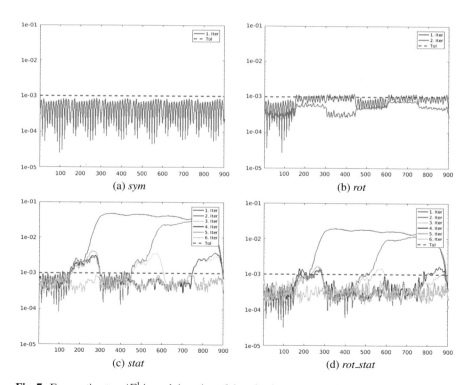

(a) *sym* (b) *rot* (c) *stat* (d) *rot_stat*

Fig. 7 Error estimator $\Delta_{\mathbf{a}}^{\mathrm{rel}}$ in each iteration of the adaptive strategy using the sets $\mathscr{K}$

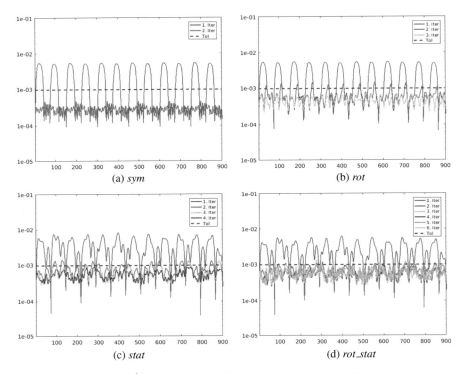

Fig. 8 Error estimator $\Delta_{\mathrm{a}}^{\mathrm{rel}}$ in each iteration of the adaptive strategy using the sets $\mathscr{M}$

A clear accuracy difference between the poles can be observed, which is reflected by the steps in the error. The settings *stat* and *rot_stat* are not handled too well by the local nature of the set $\mathscr{K}$. For each pole, a snapshot set is selected resulting in six iterations for both settings.

On the other hand, the index set $\mathscr{M}$ shows the advantages of the global nature of the snapshot selection. While for *sym* and *rot* this results in more iterations as for the set $\mathscr{K}$, for *stat* and *rot_stat* benefits can be observed. In particular for *stat*, a much faster convergence of the adaptive algorithm is obtained. The error also has a different nature. In each iteration, the error is reduced uniformly over the whole rotation and not only locally.

Next we have a look at the performance of the adaptive algorithm. For this we compare computational time (wall clock time) of the different approaches. The computational time is determined by the average over 10 runs to flatten out irregularities in the numerical realization. Note that the mesh and some constant finite element matrices that stay the same for all parameters are precomputed, and hence are not reflected in the run time. These data stay the same for all settings and are loaded from a file, and hence FEM and the adaptive POD benefit equally from it. The presented timings are the computational costs for solving (4) by each method, which is the main focus. In Tables 1 and 2, the results are summarized.

As already observed in the figures, the two index sets have different strength and weaknesses. When looking at the raw performance, the speedup can vary significantly. We get a factor of 46 for *sym* and go as low as 6 for *rot_stat* when using $\mathcal{K}$. Overall the speedup obtained by the index set $\mathcal{M}$ is better since it does not exhibit too strong variations but for *sym* and *rot*, the local nature of $\mathcal{K}$ is significantly better.

What can be observed is that although more and more snapshots are being added, the dimension of the POD basis is almost the same for the different settings. For all settings, only 13 basis vectors for the stator are required, while the number of basis vectors for the rotor depends on the problem setting. As observed during the investigation of the decay of the eigenvalues, the settings *stat* and *rot_stat* require more basis vectors.

Lastly, we have a look at the distribution of the computational time. For the FEM, we require 300–350 s to complete the simulation. Considering that 900 linear systems have been solved, the average time for solving one linear system is 0.33–0.38 s. Multiplying this with the number of computed snapshots, we can determine the overhead in the computation introduced by the basis computation, ROM generation and evaluation of the error estimator. It turns out that the overhead is between 30 and 50%. Here, efficient methods can be investigated that can update existing POD basis more efficiently. One possibility would be to build the basis hierarchical, so that in each iteration, the basis is extended and not recomputed entirely. Further, we utilized the SVD method to compute the basis vectors. This was done for stability reasons so that during the tests we do not encounter problems. Alternatively, the formulation

Table 1 Performance summary for the different settings and index set $\mathcal{K}$

Setting	FEM (s)	ROM (s)	Basis (Ω_s, Ω_r)	Iter.	Speedup	Overhead (%)
sym	328.56	7.04	(12, 6)	1	46.67	37
rot	337.17	12.76	(12, 6)	2	26.42	29
stat	341.28	46.98	(13, 16)	6	7.26	41
rot_stat	328.30	49.85	(13, 16)	6	6.58	47

Table 2 Performance summary for the different settings and index set $\mathcal{M}$

Setting	FEM (s)	ROM (s)	Basis (Ω_s, Ω_r)	Iter.	Speedup	Overhead (%)
sym	328.56	14.36	(13, 6)	2	22.88	38
rot	337.17	20.60	(13, 6)	3	16.36	34
stat	341.28	32.74	(13, 16)	4	10.42	44
rot_stat	328.30	42.17	(13, 16)	6	7.78	37

using the eigenvalue representation can be investigated [15]. This may give a boost but the overall performance will lead to similar conclusions.

Overall the numerical results are very pleasing, and the model order reduction is very effective. The speedup might not be as large when compared to strategies utilizing an online–offline decomposition. On the other hand, the presented framework can be used to directly replace the simulation routine and no further adjustments need to be done to existing code. Especially, when embedding into optimization solvers or even sampling methods like Monte Carlo simulations, this can turn out as a big benefit. Note that online–offline decompositions can be more beneficial in sampling strategies but require more initial work. The presented method is specifically intended for the speedup in optimization [16, 33]. Particularly in the robust optimization framework, the present approach is very promising [14], where ϕ and ℓ are uncertain parameters in a parametrized shape optimization problem. Further, when the ROM are utilized, a significant reduction in memory usage can be achieved. This then allows large simulation even on moderate hardware like desktop PCs.

5 Conclusion

We developed an adaptive POD snapshot sampling strategy targeted at model order reduction of electric rotating machines. A detailed description of the required component is provided. In the numerical results, different strategies of snapshot sampling were investigated and compared. The method proved to be very efficient in reducing the computational cost of symmetric and non-symmetric machines.

Acknowledgements This work is supported by the German BMBF in the context of the SIMUROM project (grant number 05M2013), by the Excellence Initiative of the German Federal and State Governments, and by the Graduate School of Computational Engineering at TU Darmstadt.

References

1. Antil, H., Heinkenschloss, M., Hoppe, R.H.W.: Domain decomposition and balanced truncation model reduction for shape optimization of the Stokes system. Optim. Method Softw. **26**(4–5), 643–669 (2011)
2. Barrault, M., Maday, Y., Nguyen, N.C., Patera, A.T.: An empirical interpolation method: application to efficient reduced-basis discretization of partial differential equations. Comptes Rendus Mathematique **339**(9), 667–672 (2004)
3. Chatterjee, A.: An introduction to the proper orthogonal decomposition. Curr. Sci. **78**, 539–575 (2000)
4. Chaturantabut, S., Sorensen, D.C.: Nonlinear model reduction via discrete empirical interpolation. SIAM J. Sci. Comput. **32**(5), 2737–2764 (2010)
5. Clénet, S.: Uncertainty quantification in computational electromagnetics: The stochastic approach. ICS Newslett. **20**(1), 2–12 (2013)
6. Davat, B., Ren, Z., Lajoie-Mazenc, M.: The movement in field modelling. IEEE Trans. Mag. **21**(6), 2296–2298 (1985)

7. Dihlmann, M., Haasdonk, B.: Certified PDE-constrained parameter optimization using reduced basis surrogate models for evolution problems. Comput. Optim. Appl. **60**, 753–787 (2015)
8. Gubisch, M., Volkwein, S.: Proper orthogonal decomposition for linear-quadratic optimal control. In: Benner, P., Cohen, A., Ohlberger, M., Willcox, K. (eds.) Model Reduction and Approximation: Theory and Algorithms (2017, to appear)
9. Haasdonk, B., Ohlberger, M.: Reduced basis method for finite volume approximations of parametrized linear evolution equations. ESAIM: M2AN. **42**(2), 277–302 (2008)
10. Henneron, T., Clénet, S.: Model order reduction applied to the numerical study of electrical motor based on POD method taking into account rotation movement. Int. J. Numer. Model. **27**(3), 1099–1204 (2014)
11. Holmes, P., Lumley, J.L., Berkooz, G., Rowley, C.W.: Turbulence, Coherent Structures Dynamical Systems and Symmetry. Cambridge University Press, Cambridge (2012)
12. Heinkenschloss, M., Sorensen, D.C., Sun, K.: Balanced truncation model reduction for a class of descriptor systems with application to the Oseen eqaution. SIAM J. Sci. Comp. **30**(2), 1038–1063 (2008)
13. Kunisch, K., Volkwein, S.: Galerkin proper orthogonal decomposition methods for parabolic problems. Numer. Math. **90**, 117–148 (2001)
14. Lass, O., Ulbrich, S.: Model order reduction techniques with a posteriori error control for nonlinear robust optimization governed by partial differential equations. SIAM J. Sci. Comput (2017, to appear)
15. Lass, O., Volkwein, S.: POD Galerkin schemes for nonlinear elliptic-parabolic systems. SIAM J. Sci. Comput. **35**(3), A1217–A1298 (2013)
16. Lass, O., Volkwein, S.: Parameter identification for nonlinear elliptic-parabolic systems with application in lithium-ion battery modeling. Comput. Optim. Appl. **62**, 217–239 (2015)
17. Monk, P.: Finite Element Methods for Maxwell's Equations. Oxford University Press, United Kingdom (2003)
18. Montier, L., Henneron, T., Clénet, S., Goursaud, B.: Transient simulation of an electrical rotating machine achieved through model order reduction. Adv. Model. Simul. Eng. Sci. **3**(1), 1–17 (2016)
19. Negri, F., Rozza, G., Manzoni, A., Quateroni, A.: Reduced basis method for parametrized elliptic optimal control problems. SIAM J. Sci. Comput. **35**(5), A2316–A2340 (2013)
20. Offermann, P., Hameyer, K.: A polynomial chaos meta-model for non-linear stochastic magnet variations. COMPEL **32**(4), 1211–1218 (2013)
21. Offermann, P., Mac, H., Nguyen, T.T., Clénet, S., De Gersem, H., Hameyer, K.: Uncertainty quantification and sensitivity analysis in electrical machines with stochastically varying machine parameters. IEEE Trans. Mag. **51**(3), 1–4 (2015)
22. Patera, A.T., Rozza, G.: Reduced Basis Approximation and A Posteriori Error Estimator for Parameterrametrized Partial Differential Equations. MIT Pappalardo Graduate Monographs in Mechanical Engineering, Boston (2006)
23. Perrin-Bit, R., Coulomb, J.L.: A three dimensional finite element mash connection for problems involving movement. IEEE Trans. Mag. **24**(3), 1920–1923 (1995)
24. Preston, T.W., Reece, A.B.J., Sangha, P.S.: Induction motor analysis by time-stepping techniques. IEEE Trans. Mag. **24**(1), 471–474 (1988)
25. Qian, E., Grepl, M., Veroy, K., Willcox, K.: A certified trust region reduced basis approach to PDE-constrained optimization. SIAM J. Sci. Comput (2017, to appear)
26. Quarteroni, A., Manzoni, A., Negri, F.: Reduced Basis Methods for Partial Differential Equations. Springer, Switzerland (2016)
27. Rozza, G., Huynh, D.B.P., Patera, A.T.: Reduced basis approximation and a posteriori error estimation for affinely parametrized elliptic coercive partial differential equations. Arch. Comput. Methods Eng. **15**, 229–275 (2008)
28. Salon, S.J.: Finite Element Analysis of Electrical Machines. Kluwer, USA (1995)
29. Shi, X., Le Menach, Y., Ducreux, J.P., Piriou, F.: Comparison of slip surface and moving band techniques for modelling movement in 3D with FEM. COMPEL **25**(1), 17–30 (2006)

30. Shimotani, T., Sato, Y., Sato, T., Igarashi, H.: Fast finite-element analysis of motors using block model order reduction. IEEE Trans. Mag. **52**(3), 1–4 (2016)
31. Sirovich, L.: Turbulence and the dynamics of coherent structures. Quart. Appl. Math. **45**(3), 561–590 (1987)
32. Toselli, A., Widlund, O.: Domain Decomposition Methods—Algorithms and Theory. Springer, Berlin (2005)
33. Zahr, M.J., Farhat, C.: Progressive construction of a parametric reduced-order model for PDE-constrained optimization. Int. J. Numer. Methods Eng. **102**, 1111–1135 (2015)

Morembs—A Model Order Reduction Package for Elastic Multibody Systems and Beyond

Jörg Fehr, Dennis Grunert, Philip Holzwarth, Benjamin Fröhlich, Nadine Walker and Peter Eberhard

Abstract Many new promising model order reduction (MOR) methods and algorithms were developed during the last decade. Industry and academic research institutions intend to test, validate, compare, and use these new promising MOR techniques with their own models. Therefore, an MOR toolbox bridging the gap between theoretical, algorithmic, and numerical developments to an end-user-oriented program, usable by non-experts, was developed called 'Model Order Reduction of Elastic Multibody Systems' (Morembs). A C++ implementation as well as a Matlab implementation including an intuitive graphical user interface is available. Import from various FE programs is possible, and the reduced elastic bodies can be exported to a variety of programs to simulate the compact models. In the course of the various projects, many improvements on the algorithmic side were added. As we learned over the years, there is not one 'optimal' MOR method. 'Optimal' MOR depends on circumstances, like boundary conditions, excitation spectra, further model usage. The toolbox is now used, e.g., in solid mechanics, biomechanics, vehicle dynamics, control of flexible structures, or crash simulations. In all these use cases, the toolbox allows the user to facilitate their well-known modeling and simulation environment. Only the critical MOR process during preprocessing is performed with Morembs, which helps to compare the various MOR techniques to find the most suited one.

1 Introduction

Numerical simulation is accepted to be the third pillar alongside theory and experiment for obtaining scientific evidence. In the future, decisions regarding design, approval, safety, and management will even more heavily depend on simulation

J. Fehr (✉) · D. Grunert · P. Holzwarth · B. Fröhlich · N. Walker · P. Eberhard
Institute of Engineering and Computational Mechanics, University of Stuttgart,
Pfaffenwaldring 9, 70569 Stuttgart, Germany
e-mail: joerg.fehr@itm.uni-stuttgart.de

P. Eberhard
e-mail: peter.eberhard@itm.uni-stuttgart.de

© Springer International Publishing AG, part of Springer Nature 2018   141
W. Keiper et al. (eds.), *Reduced-Order Modeling (ROM) for Simulation and Optimization*,
https://doi.org/10.1007/978-3-319-75319-5_7

results. Complex technical systems combine the advantages of mechanics, electronics, control, sensor information, and decision making. In addition, complex technical products often contain exchangeable subcomponents. For example, industrial robots can be equipped with diverse end effectors. In the automotive industry, the same components or even groups of components are used in different types of vehicles, such as in the Modular Transverse Toolkit (MQB) of the Volkswagen Group. The subcomponents often fulfill a functionality that can be reused in different products and, thus, save costs for a re-development. A modular design in general has many advantages. For the understanding and development of these systems, an integrated, multiphysics, and multidisciplinary view is essential. Therefore, the efficient simulation of the subcomponents and their adequate coupling are significant issues.

Let us now consider the following use case: An engineer in charge of a certain mechanical subcomponent modifies the properties of his subpart. He and his development team now need to check by simulation if the functionality, performance, and reliability of the overall system are still satisfying. A fast feedback from the simulation department is essential for the efficient further development of the technical system, e.g., a later shape optimization intended to improve the durability of the system [1].

The underlying physics of the mechanical part can often be described, e.g., by partial differential equations (PDEs). Usually, the well-established and frequently used finite element (FE) method is used to transform the PDEs into a set of ordinary differential equations (ODEs). For instance, the FE equation for solids reads

$$\boldsymbol{M}_e \cdot \ddot{\boldsymbol{q}} + \boldsymbol{k}_e(\boldsymbol{q}, t) = \boldsymbol{h}_e \tag{1}$$

with the elastic degrees of freedom $\boldsymbol{q}$, i.e., the nodal degrees of freedom, the symmetric mass matrix $\boldsymbol{M}_e \in \mathbb{R}^{N \times N}$, the elastic stiffness vector $\boldsymbol{k}_e(\boldsymbol{q}, t) \in \mathbb{R}^N$, and the applied forces $\boldsymbol{h}_e \in \mathbb{R}^N$. Possibly, the stiffness vector may depend on further quantities. This ODE system is then typically completed by suitable initial conditions. We are interested in solving the dynamical problem in the time interval $[0, t_{end}]$ to understand the behavior of the system. Nowadays, FE systems can have more than $N > 10^7$ degrees of freedom (DOFs), based on the automatic or semiautomatic meshing of 3D CAD data from a design department or 3D data from CT scans. To solve the dynamical problem, the time-discretized FE Eq. (1) needs to be solved multiple times. The computation time for one solution scales with the DOFs N of the system. Therefore, simulation studies with the full FE model are often not feasible, respectively, take to much time.

The main idea of MOR is the approximation of a large-scale dynamical system (many DOFs), e.g., the nonlinear ODE system of dimension N as given in Eq. (1), with a system of much smaller dimension n for which the most dominant features like input–output behavior, passivity, stability of the large-scale system are retained as much as possible in the small-scale system. The generation of the reduced model is mostly performed once in a so-called offline phase, which might be computationally expensive. Then, the reduced model can be simulated rapidly and possibly repeated during the so-called online phase, e.g., in the multiphysics/system-level environment.

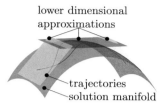

Fig. 1 The solution manifold with trajectories is locally approximated by projecton on lower dimensional, affine subspaces

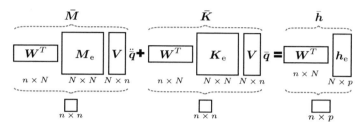

Fig. 2 Approximation of a linear FE model with a Petrov–Galerkin projection from N DOFs to n DOFs

Mainly, projection-based MOR techniques are used to approximate the state manifold by one or multiple subspaces of much lower dimension, see Fig. 1. For linear systems which are only considered in the following, primarily the Petrov–Galerkin projection is used to approximate a system on a manifold span (V) orthogonal to span(W).

As depicted in Fig. 2, a linear FE model with the original dimension N is reduced to a much smaller dimension n, where the elastic stiffness vector k_e from Eq. (1) corresponds for a linear system to $k_e = K_e \cdot q$ and all quantities with $[\bar{\ }]$ depict quantities of the reduced system. The elastic coordinates are approximated by $q \approx V \cdot \bar{q}$ with the projection matrix V, and the second projection matrix W ensures the orthogonality to the residual.

Let us summarize the use case. The engineer is responsible to derive a small-scale system, which can be simulated rapidly in the system-level environment. Model order reduction techniques are crucial to succeed in this task.

Various MOR techniques developed in the last decades, see, e.g., [2–4], can be used but have certain advantages and disadvantages. Therefore, we developed a model order reduction package, Morembs, which is capable to reduce large-scale industrial-sized problems with various MOR techniques. Furthermore, the results of the different MOR techniques can be compared to the time domain as well as in the frequency domain. Morembs, which stands for **m**odel **o**rder **r**eduction for **e**lastic **m**ultibody **s**ystems, was initiated from a need from industry to use advanced, non-modal MOR methods for the calculation of reduced bodies within elastic multibody systems (EMBS). EMBS simulations, depicted in Fig. 3, are often an appropriate

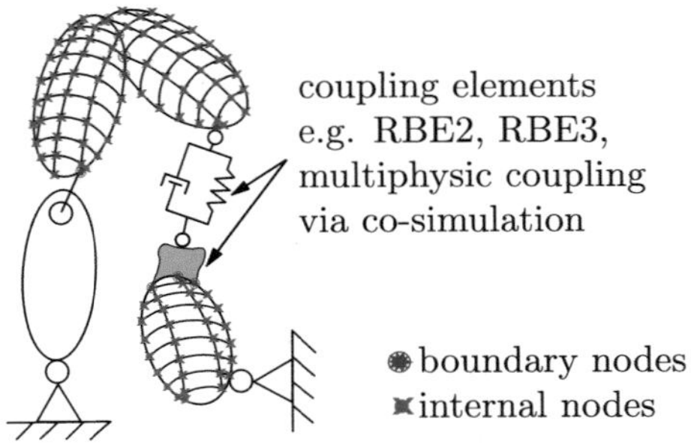

Fig. 3 Typical EMBS
system where the flexible and
rigid bodies are coupled by
joints and coupling elements

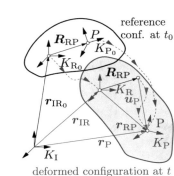

coupling elements
e.g. RBE2, RBE3,
multiphysic coupling
via co-simulation

● boundary nodes
✖ internal nodes

Fig. 4 One elastic
deformable body, the
deformation is expressed in
the floating frame of
reference formulation

reference
conf. at t_0

$\boldsymbol{R}_{RP}$ P
K_{P_0}
K_{R_0}
$\boldsymbol{R}_{RP}$
K_R
$\boldsymbol{u}_P$
$\boldsymbol{r}_{IR_0}$
$\boldsymbol{r}_{IR}$ $\boldsymbol{r}_{RP}$ P
$\boldsymbol{r}_P$ K_P
K_I

deformed configuration at t

choice to simulate mechanical subsystems in a multiphysics environment due to
their inherent modular fashion.

We refer to [5] for the advantages or disadvantages of EMBS systems. For industrial application, very often the floating frame of reference formulation, see Fig. 4, is
used. Here, the linear elastic deformation is added to the nonlinear rigid body motion
by superposition, and linear MOR techniques can be used to speed up the simulation
process.

In the following, we first give an overview about implementation details of
Morembs. It is a big advantage of Morembs that it has many interfaces to finite element package used to describe the high-dimensional model and many interfaces to
system-level simulation packages, especially EMBS software. Afterward, we shortly
describe the implemented reduction methods, how they can be classified, and what
their advantages are.

Furthermore, in this section, we speak about the algorithmic improvements
derived in the course of various projects pursued with Morembs, e.g., parametric
MOR, interface MOR to handle systems with many inputs, parametric MOR for
moving interactions as well as automated and error-controlled MOR. Finally, we
show three examples from different technical domains where Morembs was used to
derive a compact-reduced order model. These examples are a human ear, a crankshaft,
and a vehicle model. This work concludes with a summary.

2   Morembs and MatMorembs

From a practical point of view in MOR of linear FE systems or EMBS systems, three steps are typically involved in the proccess of obtaining compact and fast to calculate simulation models.

First, the original FE system, i.e., the system matrices $\{M_e, K_e\} \in \mathbb{R}^{N \times N}$, needs to be extracted. Furthermore, the interactions between the environment and the system need to be defined. Most of the times, the forces $h_e(t)$ on the system are considered as inputs $u_e(t) \in \mathbb{R}^p$ to the system distributed by the control matrix $B_e \in \mathbb{R}^{N \times p}$ and the output $y_e(t) \in \mathbb{R}^r$ captures deformations of interest via the observation matrix $C_e \in \mathbb{R}^{r \times N}$. Then, the system to be reduced is considered as a second-order multi-input multi-output (MIMO) system

$$M_e \cdot \ddot{q}_e(t) + K_e \cdot q_e(t) = B_e \cdot u_e(t),$$
$$y_e(t) = C_e \cdot q_e(t). \tag{2}$$

Second, various MOR techniques are used to calculate the projection matrices V and W for the second-order MIMO system.

Finally, a reduced second-order MIMO system

$$\bar{M} \cdot \ddot{\bar{q}}(t) + \bar{K} \cdot \bar{q}(t) = \bar{B} \cdot u_e(t),$$
$$\bar{y}_e(t) = \bar{C} \cdot \bar{q}(t) \tag{3}$$

is calculated based on the projection matrices with the reduced mass and stiffness matrices

$$\{\bar{M}, \bar{K}\} = W^T \cdot \{M_e, K_e\} \cdot V \in \mathbb{R}^{n \times n} \tag{4}$$

and the reduced control and observation matrices

$$\bar{B} = W^T \cdot B_e \in \mathbb{R}^{n \times p}, \qquad \bar{C} = C_e \cdot V \in \mathbb{R}^{r \times n}. \tag{5}$$

However, some extensions are necessary for EMBS systems. As Morembs was developed for the floating frame of reference (FFR) formulation, we elaborate on this formulation in more detail. The motion r_P of the points of the body P is split into a large nonlinearly described motion of the reference frame K_R and a linear modeled elastic motion with respect to the reference frame u_P, see Fig. 4. Therefore, one single elastic body is described with a nonlinear second-order differential equation

$$\begin{bmatrix} m I & & \text{sym.} \\ m \tilde{r}_c & J & \\ C_t & C_r & M_e \end{bmatrix} \cdot \begin{bmatrix} \ddot{q}_t \\ \ddot{q}_r \\ \ddot{q}_e \end{bmatrix} + k_c + \begin{bmatrix} 0 \\ 0 \\ K_e \cdot q_e \end{bmatrix} = \begin{bmatrix} m I \\ m \tilde{r}_c \\ C_t \end{bmatrix} \cdot g + q_A \tag{6}$$

which is divided into three parts, a translational and rotational upper part corresponding to the rigid body dynamics expressed by variables $\ddot{q}_t$ and $\ddot{q}_r$, where $\ddot{q}_t = \ddot{r}_{IR}$ is a set of Cartesian acceleration coordinates that define the origin location of the floating body reference and $\ddot{q}_r = \ddot{\Theta}$ is a set of rotational acceleration coordinates that describe the orientation of the reference frame. The lower part of Eq. (6) corresponds to the large system of elastic deformations, where q_e is the elastic coordinate which is used to approximate the elastic deformation u of body P by a Ritz approach $u(R, t) \approx \Phi(R) \cdot q_e(t)$. The subparts of the equation of motion are explained, e.g., in [5]. The deformation is usually expressed by a linear FE approach; however, this linear elastic part is coupled with C_t and C_r to the nonlinear rigid body part. In addition, the systems/bodies are coupled to the surrounding and to other bodies via the external and internal boundary conditions.

All flexible and rigid bodies of an EMBS are simulated simultaneously. To solve these large nonlinear equations efficiently, it is necessary to reduce the elastic DOFs of every subpart.

For the calculation of the projection spaces for the elastic DOFs, only the lower left part of the nonlinear ODE (6) is considered. This part of the ODE a linear, time-invariant, multi-input multi-output system, see (2), where the coupling forces due to nonlinear rigid body motion need to be considered as additional inputs, see [6].

This linear second-order MIMO system is reduced separately for every body by appropriate second-order structure preserving reduction techniques. The resulting projection spaces from the second-order MIMO system are used in Eq. (6) to calculate a reduced nonlinear ODE

$$
\begin{bmatrix} m\boldsymbol{I} & & \text{sym.} \\ m\tilde{r}_c(\bar{q}) & \boldsymbol{J}(\bar{q}) & \\ \bar{C}_t(\bar{q}) & \bar{C}_r(\bar{q}) & M_e \end{bmatrix} \cdot \begin{bmatrix} \ddot{q}_t \\ \ddot{q}_r \\ \ddot{\bar{q}} \end{bmatrix} + \bar{k}_c + \begin{bmatrix} \mathbf{0} \\ \mathbf{0} \\ \bar{K} \cdot \bar{q} \end{bmatrix} = \begin{bmatrix} m\boldsymbol{I} \\ m\tilde{r}_c(\bar{q}) \\ \bar{C}_t(\bar{q}) \end{bmatrix} \cdot g + q_A \quad (7)
$$

c body which now depends on the reduced coordinates $\bar{q}$, where the original elastic dof $q_e \approx V \cdot \bar{q}$ is approximated with the reduced elastic coordinates and $\dim(\bar{q}) \ll \dim(q_e)$. An elaborate derivation of the single terms is given, e.g., in [7]. Those reduced nonlinear ODE (7) is afterward assembled into the global EMBS and solved forward in time, see, e.g., [5].

Morembs follows this three-step approach, i.e. first import, second reduction, and finally export, see Fig. 5.

A C++ version called Morembs^{++} and a Matlab version called MatMorembs are available. The C++ implementation is based on advanced numerical libraries leading to short computation times and can be used for the reduction of industrial-sized problems with more than 1 000 000 degrees of freedom, see, e.g., [8]. The Mat-Morembs implementation is appropriate for small- to medium-sized problems and is an appropriate tool for testing and debugging new algorithms before the methods are added to Morembs^{++}. Both programs have a common data structure, and the Matlab version can be operated by directly using Matlab commands or by using an easy-to-use graphical user interface. Furthermore, MatMorembs features an automatically generated HTML documentation.

One advantage of Morembs is the existence of many interfaces to commercial FE programs. Therefore, the modeling process is performed in a well-known—and for

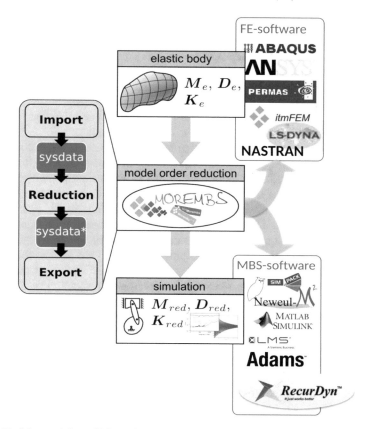

Fig. 5 Modular workflow of Morembs. First, the data of the elastic body is imported from various FE programs, available interfaces are depicted in the upper right corner. Afterward, the system is reduced, and finally, the reduced body is exported for simulation, e.g., into an MBS software. The available export interfaces are depicted in the lower right corner

the specific task most suited—FE program, as shown in Fig. 5. Currently, Morembs has direct interfaces to FEM software depicted in Fig. 5. Interfaces are available to the structural dynamic codes Ansys, Abaqus, Permas, and Nastran, but also to the crash FE codes Radioss and LS-DYNA . Furthermore, an interface to the institute-based itmFEM code exists.

After the import, the data are normalized and stored in a well-documented `struct` called `sysdata`. The second step is the model order reduction step. In this step, the user can define the requested accuracy of the reduced models, i.e., values which influence the accuracy, like number of retained modes or singular values ratio to keep, and the intended reduction method. All the data obtained during the reduction process are saved in the data structure `reddata`. Therefore, the information saved in `reddata` can be used as provenance data, to reconstruct the behavior of the reduction algorithms. Furthermore, a data structure of `sysdata` type called `redsysdata` is produced. Therefore, further reduction steps are possible by using `redsysdata` as an input to the MOR procedure. The model order reduction techniques are implemented in such a way that most features and reduction parameters are set in an automatic fashion for inexperienced users, whereas the expert user can change the behavior of the procedures by using name–value pairs.

Finally, the reduced elastic bodies are exported for standard simulation environments. The reduced matrices in (4) can also be used in Matlab.

For elastic bodies in the FFR formulation as stated before, the reduced linear MIMO system is never simulated alone. The reduced nonlinear ODE (7) for one body is described by the so-called standard input data (SID) format [9] which contains a Taylor approximation of the nonlinear mass invariants. Once the mass invariants are calculated, they can be written into various output formats readable by standard EMBS programs, e.g., Simpack, LMS Virtual.Lab, Recurdyn, Adams, First, or Neweul-M^2. Another output possibility is the output of the reduced system matrices in various data formats, e.g., as a Simulink state-space block or as the standard LTI system description (A, B, C, D) or written in an ASCII format readable by AMESim. Finally, it is possible to export the calculated projection matrices V or W into a format readable by standard FE programs. The standard FE program then treats the calculated projection matrices V similar to self-computed eigenmodes. The converter, export functions, and substructure functions of the FE programs can be used to export the non-modal reduced models to various other simulation programs. Alternatively, the non-modal reduced models can be used in the FE program as a substructure.

We want to summarize this section and the description of the main properties Morembs. It should be finally mentioned again that the modeling and simulation happen in the familiar simulation environment, and only the MOR step is made with Morembs.

Further options of Morembs are:

- Calculation of stress modes, which can be used in an EMBS environment to calculate the stress distribution online, e.g., for durability-based examinations [10, 11].
- Visualization of the imported FE models and ansatz spaces
- Usage of reduced optical bodies within the coupled dynamical–optical simulation environment OM-SIM, see, e.g., [12].
- Coupling with Famous [13] for uncertainty investigation for elastic multibody systems, compare, e.g., [14].
- Usage of multiple analysis tools to compare different reduction methods, like the MAC criteria or the relative error measured in the $\mathscr{H}_2$ or $\mathscr{H}_{inf}$ norm.

3 Implemented MOR Methods and Algorithmic Improvement

During the last decades, various MOR methods were developed. An overview for linear methods is given in [15]. Most of the methods were developed for first-order state-space systems. For MOR of elastic multibody systems and for MOR of linear FE models, it is advantageous to keep the second-order structure of the system [5]. Therefore, Morembs keeps the second-order structure. Most of the MOR techniques are adaptively programmed, meaning that if a second-order structure is an input to the program, then second-order reduction techniques are used, whereas if a first-order system is given, then first-order reduction techniques are used automatically.

In the course of various publicly and industrially funded projects, the following major research questions were asked and partly answered:

- How to couple different reduced order models with each other? What system properties are retained when multiple reduced order models are interconnected with each other?
- What happens for systems with non-stationary boundary conditions?
- How is an automated and error-controlled model order reduction process possible?
- What happens if the system has many inputs?

Therefore, in the following, we try to give an overview about some major results achieved with Morembs and try to answer the previously defined research questions.

Model Reduction Classes and Coupling of Reduced Order Models

We will distinguish between different classes of MOR methods, some of which can retain certain properties, e.g., passivity or stability. We distinguish between the following MOR classes:

- truncation-based methods
  - modal truncation
  - balanced truncation
- interpolatory based
  - Guyan condensation
  - Krylov subspace-based methods
- combination of both methods

Truncation-Based Methods

For truncation-based MOR methods, the second-order system is transformed with a transformation matrix $\boldsymbol{\Phi}$ into a representation for which a specific importance can be identified for each state. Afterward, the less important states are truncated from the system.

The best-known truncation method is modal reduction, for which the system (2) is transformed by its eigenspace into its eigenspace representation. Therefore, the homogenous eigenproblem $(\lambda_i M_e + K_e) \cdot \phi_i = 0$ is solved to calculate the eigenmodes ϕ_i belonging to the eigenvalue λ_j. The excitation of the system is not considered, which is an advantage and disadvantage. If the excitation of the system is not known a priori or is subject to variations, modal reduction can be very suitable. However, even if the inputs to the system, e.g., the frequency content of the excitations, are well-known, this information cannot be used in the reduction process, and therefore, modally reduced bodies are usually not as good as other MOR techniques which explicitly use this information.

Therefore, balanced truncation methods are popular, compare, e.g., [2]. Here, the input-to-output map is explicitly used in the reduction process, and they often even have a priori error bounds. Those states not much involved in the energy transfer from input to output are truncated. The controllability and observability Gramian matrices, P and Q, and the Hankel singular values $\sigma_k = \sqrt{\lambda_k(P \cdot Q)}$ are strongly related to balanced truncation techniques. They identify those states which are simultaneously easy to control and easy to observe. Furthermore, frequency weighting is possible. For second-order systems, usually the projection is made on the dominant subspace of the frequency-weighted position Gramian matrix P^p, see [5] or [16] how to calculate those dominant subspaces by a greedy-based approximation of the frequency-weighted Gramian matrices by a POD approach.

Unfortunately, there is not one optimal MOR method. Which method is most suited always depends on the use case. For example, balanced truncation-based MOR has rigorous error bounds, which can be tuned in a certain frequency range and lead to very good reduction results if we reduce only one body. However, error bounds that are guaranteed for balanced truncation methods do not hold anymore for connected system consisting of multiple reduced subcomponents after the introduction of connections to the environment as explained in [6].

Interpolatory-Based Methods

The second class of MOR methods are interpolatory-based methods. The transfer function $H(s) = C_e \cdot (s^2 M_e + K_e)^{-1} \cdot B_e$ in the complex s-domain of the system is interpolated by the reduced order model—with the reduced transfer function $\bar{H}(s) = \bar{C} \cdot (s^2 \bar{M} + \bar{K})^{-1} \cdot \bar{B}$—at certain expansion point σ_0 via moment matching.

The so-called moments $T_j^{\sigma_0}$ of order j around the shift σ_0 of the system can be used to express a transfer function $H(s)$ with a power series

$$H(s) = \sum_{j=0}^{\infty} -T_j^{\sigma_0} s^j = \sum_{j=0}^{\infty} -\frac{1}{j!} \frac{\partial^j H(s)}{\partial s^j} s^j . \tag{8}$$

With the help of projections on Krylov subspaces, numerically robust implementations are available to impose that certain moments of the original transfer function

$H(s)$ and the reduced transfer function $\bar{H}(s)$ match. For example, if the zeroth moment matches between original and reduced system, the transfer function developed around the expansion point σ_0 is the same for $H(s)$ and $\bar{H}(s)$. If the first-order moments are matching, the derivatives of $H(s)$ and $\bar{H}(s_k)$ are the same. One concern in MOR with Krylov subspaces is the correct selection of the expansion points and error bounds. Various approaches exist, where the iterative rational Krylov algorithm (IRKA) [17] is one of the most popular. A more recent work about error bounds and an automatic choice of parameters is, e.g., [18].

One big advantage for connected systems is that the moment matching conditions which are introduced with interpolation methods for single components also hold for an assembled system [6]. The most prominent existing methods for MOR in structural mechanics are component mode synthesis methods, which usually combine different approaches, see, e.g., [19]. Usually, interpolation-based modes like constraint modes (Craig–Bampton) or attachment modes (Rubin–McNeal) are combined with certain eigenmodes of the system. Vast improvement can be achieved if second-order balanced truncation methods are used for the internal dynamics instead of eigenmodes, see [20]. For this new class of MOR methods, it is crucial that new inputs are defined for the internal dynamics during the input–output reduction process of the internal dynamics because the acceleration of the boundary degrees of freedom applies inertia forces onto the internal dynamics of the systems.

Non-stationary Boundary Conditions

In a rising number of engineering applications, the system matrices describing the elastic body are not constant but parameter dependent. Such kinds of systems arise, for example, in the simulation of gear trains, sliding components, turning and milling processes, or structural optimization. The equation of motion for such parametric systems then becomes

$$
\begin{aligned}
M_e(p) \cdot \ddot{q}_e(t) + K_e(p) \cdot q_e(t) &= B_e(p) \cdot u_e(t), \\
y_e(t) &= C_e(p) \cdot q_e(t).
\end{aligned}
\tag{9}
$$

Since the input–output behavior of such systems is obviously parameter dependent, the use of standard input–output-based MOR methods for linear time-invariant systems yields only very poor results. A summary of reduction methods for parametric systems can be found in [21]. There, a distinction between local and global approaches is suggested. In local approaches as in [22] or [23], the parametric system is reduced for different individual parameter values. Then, after a congruence transformation, the individually reduced system matrices are interpolated. In contrast, global approaches as described in [24] try to find one representative subspace capturing the entire parameter-dependent system dynamics. This usually results in a

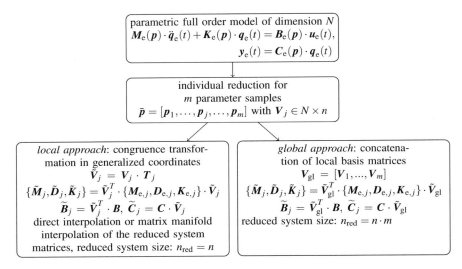

Fig. 6 Reduction process for local and global parametric model order reduction

larger reduced size of the reduced model. The reduction process for global and local approaches is shown in Fig. 6.

Matmorembs provides efficient data structures for parametric systems, interpolation routines for multidimensional parameter spaces, and tools for a priori and a posteriori error analysis. The local approach from [22] is adopted to retain special block matrix structures of the reduced parametric systems in order to ensure numerically efficient evaluations of the equation of motion. For global parametric model order reduction approaches, an intelligent sampling of the parameter space and advanced reduction strategies are necessary to keep the order of the reduced model moderate. First, several types of IRKA for second-order systems as described in [25] are available for matching the transfer function (and its gradient) of the parametric reduced order model with the transfer function of the full-order parametric model. Second, the error estimator from [16] is adapted for parametric systems. With that, the error estimator can be used in a greedy procedure in order to minimize the approximation error both over the parameter space and in a limited frequency interval, providing an automated, adaptive reduction process for parametric systems. For both the local and the global approaches, interfaces are available to incorporate the parametrically reduced bodies in the simulation of elastic multibody systems. For further information about the advantages and disadvantages of these techniques, especially for the application in elastic multibody systems, we refer to [26, 27] or [28].

Automated and Error-Controlled MOR

As far as safety questions are concerned, knowledge about the error introduced by reduction is crucial. The importance of error estimation for MOR has led to various error estimation techniques. Very often, the error is measured in the frequency domain or in a specific system norm [18]. A priori error bounds give statements of the quality of the reduced model based on the offline construction procedure. For example, in SVD-based techniques, the worst-case behavior of the reduced models can be estimated by determining the neglected (Hankel) singular values [2, 5]. The error here is measured in frequency space. Additionally, error approximations are available for Krylov subspace reduction methods, see, e.g., [5, 18], to select the most suitable expansion points. Alternatively, the error can be measured directly in the time domain. It is no surprise that L2-error estimates in state and frequency space are connected by Parseval-type equalities [15].

While a priori error bounds give worst-case behavior bounds but ensure good approximation independent of the simulation setting, the individual simulation runs could be much better than this worst case. This means that the a priori error bound might be largely overestimating the actual error. This leads to the idea or require-ment of a posteriori error estimation of reduced systems: For each special input signal, loading case, parameter, etc., the reduced model should give additional error information for its current simulation setting. This was always the focus of certified reduced basis methods, see, e.g., [29]. A major advantage of the a posteriori error estimator is that the estimator is independent of the used reduction technique. For time-dependent problems, the error estimates are obtained by accumulating residual norms over time with suitable multiplication of growth factors [29]. The solution of this augmented reduced system then results in an additional system component which represents the error bound. Therefore, in [30], the a posteriori error estimator for linear first-order systems, see [29], is extended for error estimation of mechan-ical second-order systems. Due to the special second-order structure of mechanical systems, an improvement of the a posteriori error estimator is achieved.

Many Inputs Case

In development processes, different components of a technical product are often developed and modeled in different departments. To obtain a model of the assembled system, the component models have to be connected in a separate step. For the reduc-tion of the component models, interpolation-based methods are advantageous, see [6]. To obtain properly reduced models, the degrees of freedom that are constrained by this connection have to be defined as inputs for the model order reduction step, see [31]. If the component models are coupled via large surfaces, like, for example, a gear wheel on a drive shaft, the model will have a large number of inputs. When the model is reduced using interpolation-based methods, this leads to large reduced

Fig. 7 Reduction process
using interface reduction

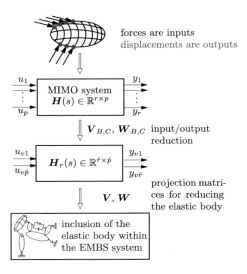

models, as the number of inputs gives a lower bound for the achievable reduced
system dimension. To overcome this problem, interface reduction can be used, for
example, a Singular Value Decomposition Model Order Reduction (SVDMOR) or
an Extended SVDMOR called ESVDMOR, see [20, 32].

Based on the input/output behavior of the system, the dominant input and output
directions are determined. For example, the most important deformation patterns or
the eigenmodes of the input space are used, see, e.g., [6, 33]. This leads to a reduced
input description. Therefore, smaller reduced models using interpolation-based methods can be obtained. The interface reduction fits well in the reduction process and
does not influence the processing of the elastic multibody system, compare Fig. 7.

4 Use Cases with Morembs

In the following section, we will give three use cases where we used Morembs
to derive reduced order models. The first example is a biomechanical example—
the human middle ear; the second example is a classical mechanical engineering
example—a crankshaft of an engine is simulated in a multiphysics environment; and
the last example is a racing kart frame used in noise, vibration, and harshness (NVH)
and crash simulations.

The Middle Ear as an Elastic Multibody Systems

The human ear is a complex acoustic analyzer for the identification and classification of sound events. The function of pinna, ear canal, and middle ear is the transfer of sound events from the free field to the inner ear, see Fig. 8.

For the better understanding of the human ear, to assist surgeons, and for the development of new active and passive prostheses, various computer-based experiments are conducted. EMBS simulations are well suited to simulate the behavior of the middle ear. The tympanic membrane, the air in the ear canal, and the tympanic (middle ear) cavity are considered as elastic bodies. They are first modeled using the finite element method. The large number of degrees of freedom makes a following reduction step of the acousto-structural finite element model inevitable. The problem to be solved is an acousto-structural coupled problem, which is modeled using the pressure-based Eulerian approach. As stated in [34], this leads to a coupled system

$$\begin{bmatrix} \boldsymbol{M}_{\mathrm{s}} & \boldsymbol{0} \\ \rho_0 \cdot \boldsymbol{R}^T & \boldsymbol{M}_{\mathrm{f}} \end{bmatrix} \cdot \begin{bmatrix} \ddot{\boldsymbol{q}}_{\mathrm{e}} \\ \ddot{\boldsymbol{p}} \end{bmatrix} + \begin{bmatrix} \boldsymbol{K}_{\mathrm{s}} & \boldsymbol{R} \\ \boldsymbol{0} & \boldsymbol{K}_{\mathrm{f}} \end{bmatrix} \cdot \begin{bmatrix} \boldsymbol{q}_{\mathrm{e}} \\ \boldsymbol{p} \end{bmatrix} = \begin{bmatrix} \boldsymbol{F}_{\mathrm{s}} \\ \boldsymbol{F}_{\mathrm{f}} \end{bmatrix} \qquad (10)$$

with asymmetric system matrices. Due to the asymmetric system matrices, the system has distinct left and right eigenvectors. For the preservation of stability, oblique modal reduction is used.

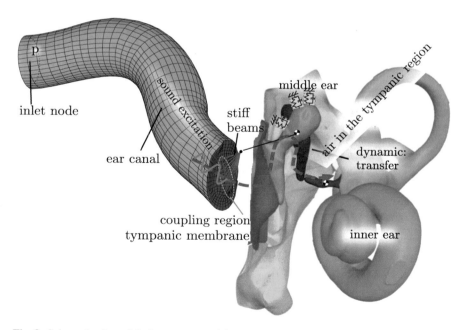

Fig. 8 Schematic view of the human ear model

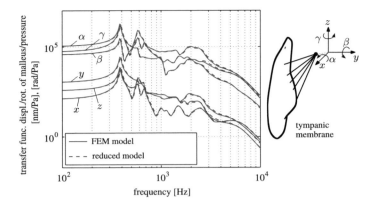

Fig. 9 Comparison of the transfer functions of the FEM model (solid) and the reduced model (dashed). The input is the pressure in the ear canal, and the output is the spatial motion of the interface node

By means of model order reduction, the local models with initially 16 022 degrees of freedom are reduced to the size of 40 degrees of freedom. In Fig. 9, the transfer behavior of the FEM model—the input is the pressure in the ear canal and the outputs are translational and rotational motions of the interface nodes, i.e., the center of gravity of the malleus—is compared with the reduced order model. Both the rotational displacement and translational displacement of the reduced model are in good accordance with the full FEM model.

To represent the nonlinearity of the tympanic membrane in the reduced model, a parametric model order reduction approach is used. It is assumed that the nonlinearity is represented by the relative pressure between the ear canal and the tympanic cavity. Therefore, matrix interpolation is used, and the results were satisfactory, see [34]. Therefore, a reduced order model included in the EMBS model is suitable to represent the dynamical behavior of this biomechanical system. Over the last years, the EMBS model of the ear was even further improved, see, e.g., [35]. For example, the contact interactions of the rigid ossicular chain are improved. Here, again MatMorembs is helpful. The high-resolution FE model of the ossicular chain, derived from micro-CT data, is translated via MatMorembs into a EMBS in Neweul-M^2. Especially, the contact surfaces are needed for the calculation of the contact forces, which have a large influence on the dynamic behavior.

Flexible Crankshaft of a Combustion Engine

Our second example is a typical mechanical engineering example. However, this mechanical part is just a subpart of a multiphysics system. The crankshaft as a part of the crank drive is in charge of transforming the forces from the piston rods into a

Fig. 10 Structure of a crank drive

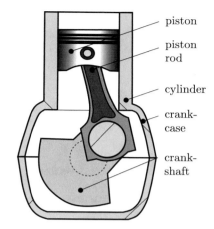

piston

piston rod

cylinder

crank-case

crank-shaft

Fig. 11 Crankshaft of a four-cylinder engine with crank webs, main bearing, and pin bearing

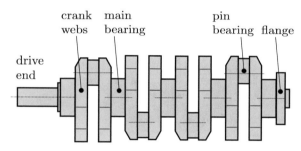

crank webs   main bearing   pin bearing   flange

drive end

rotational motion, see Fig. 10. To improve the durability of the engine and optimize the overall behavior, e.g., for fuel efficiency and emission reduction, simulations are very important. The major parts of the engine, depicted in Fig. 10, are the pistons, the piston rods, the crankshaft with the counterbalance like the flywheel. The bearings are also very important for the systems. They connect the crankshaft see Figs. 11 and 12, to the piston rods and the engine block via a fluid film lubrication. The mechanical parts are simulated within the elastic multibody system code Simpack [36]. Input to the system is the gas forces, which act on the pistons. The tribological behavior, i.e., the hydrodynamic interaction, e.g., at the lubrication gap, is simulated within FIRST [37]. For a correct calculation of the lubrication geometry, the consideration of the elasticity of the components is very important. In Fig. 13, a comparison is drawn between the maximum pressure inside the bearings concerning the usage of a rigid and an elastic crankshaft. The difference makes it obvious that an elastic description is required.

Among the challenges for the simulation of such a multiphysics systems is the large number of interfaces. Therefore, suitable interface modeling strategies, respectively interface approximation techniques, are very important, see, e.g., [20, 32, 33]. In order not to pay attention to every node on the surface of the bearing cones in the crankshaft example, the coupling of the crankshaft is designed by interface

Fig. 12 Meshed CAD model of a crankshaft

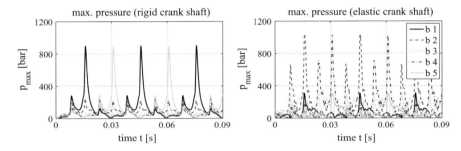

Fig. 13 Maximum pressure inside the five main hydrodynamic bearings, b1 to b5 (left: rigid, right: elastic) crankshaft

nodes, see [38]. Overall, 35 inputs and outputs are defined: each radial, translatory degree of freedom at the pin bearings, every translatory degree of freedom at the main bearings, and every degree of freedom at the drive end and the flange. By an analysis of the actor signal and the torque caused by the flywheel and the torsional damper, the interesting frequency range from $f = 0$ Hz to about $f = 720$ Hz is identified. The frequency information is helpful to tune certain reduction methods, e.g., the frequency-weighted Gramian matrix methods or Krylov subspace reduction methods. In Fig. 14, the relative approximation error

$$\varepsilon\left(f\right) = \frac{\|\boldsymbol{H}\left(f\right) - \overline{\boldsymbol{H}}\left(f\right)\|_F}{\|\boldsymbol{H}\left(f\right)\|_F} \tag{11}$$

in the Frobenius norm of different reduction methods with the same reduced size of 70 is shown. In this example, non-modal reduction methods are superior because the input–output map is known a priori and is explicitly used in the reduction process. Furthermore, only one part of the whole system is reduced and not an interconnected system. Therefore, Fig. 14 represents the typical behavior of the different reduction methods if we only look at the single component.

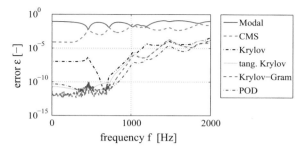

Fig. 14 Relative error of different reductions with the same reduced size of 70

In addition, dynamic simulations are considered and the different reduction results are compared with respect to accuracy in the time domain. As reference for the dynamical simulation, a crankshaft reduced to a quite high number of 140 ansatz functions is taken. As tribological quantities, the minimum gap and the maximum pressure in the bearing next to the flange of the flywheel are displayed. The relative error of the tribological quantities, the output's minimum gap, and maximum pressure

$$\varepsilon\left(t\right) = \frac{|\boldsymbol{y} - \overline{\boldsymbol{y}}|}{|\boldsymbol{y}|} \tag{12}$$

are depicted in Fig. 15. The potential of modern reduction techniques is shown by a comparisons of different reductions to the same reduced size. Substantial benefits in the relative error can be gained for a single elastic body. However, the interface reduction strategy is crucial. In the time simulation, no differences concerning the tribological quantities can be recognized, the minimum gap and the maximum pressure when comparing different reduced crankshafts with a comparable error in the frequency domain. So in this context, it can be said that the usage of model order reduction is very suitable to achieve a good tribological consistency within a multiphysics simulation environment.

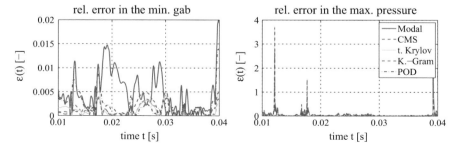

Fig. 15 Relative error of the results of the dynamic simulation for a short period

Racing Kart Frame Used in NVH Simulations and Crash Applications

The third example is an example from vehicle dynamics, respectively vehicle safety. Depicted in Fig. 16 is the frame of a racing kart. The racing kart is a simple automobile, and the dynamic behavior of the racing kart is strongly influenced by the structural characteristics of the elastic frame. In [39], the influence of MOR on the noise, vibration, and harshness (NVH) relevant states is discussed. The large-scale model of the kart frame is modeled in Ansys. The reduction takes place in MatMorembs, and afterward, the different reduced models are exported to either Simpack or Neweul-M^2 to analyze the NVH relevant states of a vehicle, e.g., the ride characteristics if we drive over Belgian block.

However, another important research field is the acceleration of crash simulations. Nowadays, crash simulation with commercial finite element (FE) software is a core area of vehicle development. Bearing in mind that crash simulation is among the most calculation time-consuming tasks in car design, the usage of model order reduction for speed up and data reduction is a logical consequence. Likewise, the racing kart's crash behavior is strongly influenced by the nonlinear plastic deformation characteristics of the frame. Therefore, the kart frame is an ideal example to discuss the performance of various model order reduction methods for structures in crashes, see [33]. The racing kart crashing into a rigid pole and exhibiting a plastic deformation should be represented by a reduced order model in LS-DYNA. Consequently, the capabilities of MatMorembs were extended to work with the explicit FE code LS-DYNA.

The workflow of the MOR approach in crash application is depicted in Fig. 17 and explained in detail in [33]. In a first step, parts with linear and nonlinear characteristics are identified. In order to determine which parts of the model exhibit small deformations and, therefore, can be described with a linear model, it is necessary to run multiple simulations with varying parameters, e.g., impact velocity in an offline step, see the upper left part in Fig. 17 and [33]. Afterward, the identification of linear and nonlinear behavior as depicted in the upper middle part in Fig. 17

Fig. 16 Racing kart used as a surrogate model for a real car

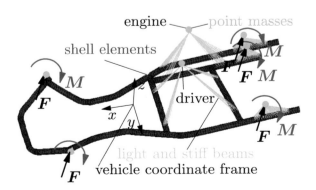

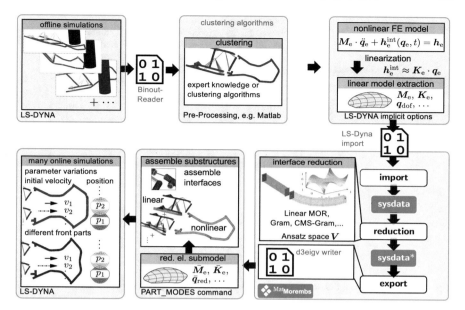

Fig. 17 Workflow of the process suggested to speed up explicit simulations. The workflow shows how MatMorembs is used to calculate the reduced elastic body for the online simulation

is performed either by an experienced user or by the application of methods from machine learning, see [40] or [41]. In the next step, the model is substructured into nonlinear and into linear parts. Our focus is on the linear part which is approximated by linear ansatz functions. Accordingly, only the parts marked as linear are linearized and exported by LS-DYNA. Afterward, the linearized LS-DYNA model is imported into MatMorembs. As a consequence, all the non-modal and additional features of MatMorembs are now available and can be used to calculate an optimal ansatz space.

Also for this example, the correct modeling and handling of the interfaces between the linearly approximated and nonlinear parts is critical for good reduction results [33]. The various interface reduction methods available in MatMorembs which are mentioned in Sect. 3 can be applied. For this model, either the local 'modal' deformation pattern at the single interface or the global 'deformation' pattern of all interfaces is used to reduce the number of interfaces. Subsequent to the interface reduction, the most suitable ansatz functions to approximate the rear part are calculated. Afterward, the linearized reduced model is exported toward LS-DYNA, see the lower right part of Fig. 17. LS-DYNA assembles a reduced elastic body with the *PART_MODES command, where within LS-DYNA a floating frame of reference approach is used to describe reduced elastic bodies. Finally, the reduced model consisting of a linear reduced part and a nonlinear part is simulated with LS-DYNA in the online step. The reduced model can be used to perform many optimization or parametric runs, see the lower left part of Fig. 17. In Fig. 18, the comparison between the full and reduced model is depicted for the central crash with 32 km/h. With the

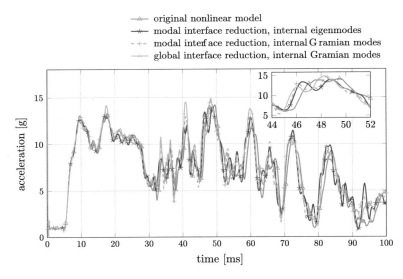

Fig. 18 Acceleration of the driver calculated with different reduction approaches

reduced models, we achieved a good match until 30 ms; after 30 ms, we have a small phase shift. Mainly, the timing of the peaks varies; for the original model, the main acceleration peaks occur later. The results of the different reduction methods—all models have 6 rigid body degrees of freedom plus 51 elastic degrees of freedom— are very similar. The size of the reduced system was selected based on the decay rate of the Hankel singular values of the internal Gramian matrix modes. For this reduction size, there is not much difference between the usage of internal Gramian matrix modes and the usage of internal normal modes.

In addition, the plastic strain of the reduced, using CMS Gram with modal interface reduction, nonlinear model is calculated and depicted in Fig. 19. The maximum plastic strain is 61%, which is very similar to the original strain, and the maximum strain occurs at the same positions as for the original model, see [33]. The small difference in the kinematic behavior between the full and the reduced models can be seen if we compare the final state of the original model which is also depicted in Fig. 19. The biggest challenge in this example was the coupling between reduced and original model. Without a suitable interface reduction method, no simulation of the reduced model was possible. However, the runtime improvement of the reduced model in comparison with the original model is rather disappointing. The advantage of fewer and therefore faster element processing, resulting in a calculation time of $0.97 \cdot 10^3$ s versus $2.21 \cdot 10^3$ s, is devoured by the time needed to calculate the reduced elastic body ($1.26 \cdot 10^3$ s). The simulation time of the reduced elastic body should be much faster than the current implementation in LS-DYNA, from the experience gained in other projects. However, this is not the focus of our research and of Morembs. The

Fig. 19 Plastic strain of the deformed, reduced model (CMS Gram with modal interface reduction) plus the kinematics of the unreduced model (gray) for comparison

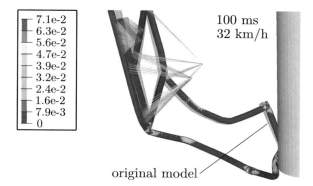

7.1e-2
6.3e-2
5.6e-2
4.7e-2
3.9e-2
3.2e-2
2.4e-2
1.6e-2
7.9e-3
0

100 ms
32 km/h

original model

focus of the development of Morembs is the possibility to use standard modeling tools and standard simulation tools. Using advanced and elaborate MOR techniques to connect the different simulation domains was achieved with this example.

5 Conclusions and Challenges

Nowadays, we have a modular setup of technical products. Therefore, the modular simulation approach in EMBS is very suitable in the development process of new technical systems. MOR techniques offer the possibility to create data bases of reduced bodies which can be used, e.g., for real-time simulations, within optimization loops or in multiquery simulation scenarios. Morembs is a powerful toolbox for the analysis and reduction of large-scale dynamical models. It is the workhorse at our institute for MOR in EMBS but has also many interfaces to different other codes. The advantage of Morembs is the usage of standard FE programs and standard MKS programs for modeling and simulation which preserves the familiar process chain. However, we can use modern reduction methods instead of only modal methods and have further analysis capabilities. Usually, those input-/output-based MOR delivers more reliable and predictable results, which is very important for automation. They usually deliver better approximations, faster convergence rates, and shorter computation times. Interpolation-based methods are advantageous for coupled system, because they preserve some system properties. The well-known CMS-based MOR approaches are improved by combining them with Gramian matrix-based approaches [42]. Therefore, it is crucial to consider the inertia-based forces of the boundary as inputs on the internal dynamics.

Taking into account the time-varying boundary conditions during the MOR process is crucial for the approximation quality, e.g., by an interface approximation. Very often, the interface approximation strategy is more crucial than the MOR strategy. Furthermore, parametric model order reduction is one suitable solution for moving interaction forces and is implemented in Morembs.

Code Availability / Licensing Option
Morembs was used to compute the presented results and further information can be obtained at:

http://www.itm.uni-stuttgart.de/research/model_reduction/MOREMBS_en.php

It is authored by the Institute of Engineering and Computational Mechanics, University of Stuttgart, Pfaffenwaldring 9, 70569 Stuttgart, Germany
Please contact Prof. Peter Eberhard [peter.eberhard@itm.uni-stuttgart.de] for licensing information.

Fig. 20 Code availability

Furthermore, the software Morembs is also suitable to be used in other use cases due to its modularity, e.g., for the acceleration of explicit crash simulations or in opto-mechanical simulations. More information about the availability of the program is given in Fig. 20.

Current topics and challenges are alternative modeling of interactions, the development of nonlinear and non-intrusive MOR techniques, and the time-domain error estimation for nonlinear systems.

Acknowledgements The authors would like to thank the German Research Foundation (DFG) for financial support of the project within the Cluster of Excellence in Simulation Technology (EXC 310/1) at the University of Stuttgart, the support of this research work within the project EB 195/11-1, FE 1583/2-1, Ei 231/6-1, the FVV (Forschungsvereinigung Verbrennungskraftmaschinen e.V.) with its working groups 'Optimale FE Reduktion' and the Automotive Simulation Center Stuttgart (ASCS) for providing partially the funding for this research.

References

1. Tobias, C., Fehr, J., Eberhard, P.: Durability-based structural optimization with reduced elastic multibody systems. In: Proceedings of the 2nd International Conference on Engineering Optimization, Paper-ID 1119, Lisbon, Portugal, 6–9 Sept 2010
2. Antoulas, A.: Approximation of Large-Scale Dynamical Systems. SIAM, Philadelphia (2005)
3. Benner, P., Mehrmann, V., Sorensen, D. (eds.): Dimension Reduction of Large-Scale Systems. Lecture Notes in Computational Science and Engineering, vol. 45. Springer, Berlin (2005)
4. Schilders, W., van der Vorst, H., Rommes, J. (eds.): Model Order Reduction: Theory, Research Aspects and Applications. Mathematics in Industry, vol. 13. Springer, Berlin (2008)
5. Fehr, J.: Automated and error-controlled model reduction in elastic multibody systems. Dissertation, Schriften aus dem Institut für Technische und Numerische Mechanik der Universität Stuttgart, vol. 21. Shaker Verlag, Aachen (2011)
6. Holzwarth, P., Eberhard, P.: Interpolation and truncation model reduction techniques in coupled elastic multibody systems. In: Proceedings of the ECCOMAS Thematic Conference on Multibody Dynamics, Barcelona, Spain (2015)
7. Schwertassek, R., Wallrapp, O.: Dynamik Flexibler Mehrkörpersysteme (in German). Vieweg, Braunschweig (1999)
8. Volzer, T., Eberhard, P.: Arch. Mech. Eng. **63**(4) (2016)

9. Wallrapp, O.: Standardization of flexible body modeling in multibody system codes. Part I: definition of standard input data. Mech. Struct. Mach. **22**(3), 283–304 (1994)
10. Tobias, C., Matha, D., Eberhard, P.: Durability-based data reduction for multibody system results and its applications. Euromech Newsl. **40**, 15–24 (2011)
11. Tobias, C.: Schädigungsberechnung in elastischen Mehrkörpersystemen (in German). Dissertation, Schriften aus dem Institut für Technische und Numerische Mechanik der Universität Stuttgart, vol. 24. Shaker Verlag, Aachen (2012)
12. Wengert, N.: Gekoppelte dynamisch-optische Simulation von Hochleistungsobjektiven (in German). Dissertation, Schriften aus dem Institut für Technische und Numerische Mechanik der Universität Stuttgart, vol. 40. Shaker Verlag, Aachen (2015)
13. Hanss, M., Walz, N.P.: Software FAMOUS. Universität Stuttgart, Institut für Technische und Numerische Mechanik (2015). http://www.itm.uni-stuttgart.de/research/famous. Accessed 08 May 2017
14. Iroz, I., Hanss, M., Eberhard, P.: Simulation of friction-induced vibrations using elastic multibody models. In: Proceedings of the ECCOMAS Thematic Conference on Multibody Dynamics, Barcelona, Spain (2015)
15. Antoulas, A.C.: On Pole Placement in Model Reduction, at -. Automatisierungstechnik **55**(9), 443–448 (2007)
16. Fehr, J., Fischer, M., Haasdonk, B., Eberhard, P.: Greedy-based approximation of frequency-weighted gramian matrices for model reduction in multibody dynamics. Zeitschrift für Angewandte Math. Mech. **93**(8), 501–519 (2012)
17. Gugercin, S., Antoulas, A.C., Beattie, C.A.: $\mathscr{H}_2$ model reduction for large-scale linear dynamical systems. SIAM J. Matrix Anal. Appl. **30**(2), 609–638 (2008)
18. Panzer, H.K.F.: Model order reduction by Krylov subspace methods with global error bounds and automatic choice of parameters. Dissertation, Technische Universität München. Verlag Dr. Hut, München (2014)
19. Craig, R.: Coupling of substructures for dynamic analyses: an overview. In: Proceedings of the AIAA Dynamics Specialists Conference, Atlanta (2000)
20. Holzwarth, P., Eberhard, P.: Interface reduction for CMS methods and alternative model order reduction. In: Proceedings of the MATHMOD 2015–8th Vienna International Conference on Mathematical Modelling, Vienna, Austria (2015)
21. Benner, P., Gugercin, S., Willcox, K.: A survey of projection-based model reduction methods for parametric dynamical systems. SIAM Rev. **57**(4), 483–531 (2015)
22. Panzer, H., Mohring, J., Eid, R., Lohmann, B.: Parametric model order reduction by matrix interpolation, at -. Automatisierungstechnik **58**(8), 475–484 (2010)
23. Amsallem, D., Farhat, C.: An online method for interpolating linear parametric reduced-order models. SIAM J. Sci. Comput. **33**, 2169–2198 (2011)
24. Baur, U., Beattie, C., Benner, P., Gugercin, S.: Interpolatory projection methods for parameterized model reduction. SIAM J. Sci. Comput. **33**, 2489–2518 (2009)
25. Wyatt, S.: Issues in interpolatory model reduction: inexact solves, second-order systems and DAEs. Dissertation, Virginia Polytechnic Institute and State University, Blacksburg, 2012
26. Fischer, M., Eberhard, P.: Simulation of moving loads in elastic multibody systems with parametric model reduction techniques. Arch. Mech. Eng. **61**(2), 209–226 (2014)
27. Fischer, M., Eberhard, P.: Application of parametric model reduction with matrix interpolation for simulation of moving loads in elastic multibody systems. Adv. Comput. Math., 1–24 (2014)
28. Baumann, M.: Parametrische Modellreduktion in elastischen Mehrkörpersystemen (in German). Dissertation, Schriften aus dem Institut für Technische und Numerische Mechanik der Universität Stuttgart, vol. 43. Shaker Verlag, Aachen (2016)
29. Haasdonk, B., Ohlberger, M.: Efficient reduced models and a-posteriori error estimation for parametrized dynamical systems by offline/online decomposition. Math. Comput. Model. Dyn. Syst. **17**(2), 145–161 (2011)
30. Ruiner, T., Fehr, J., Haasdonk, B., Eberhard, P.: A-posteriori error estimation for second order mechanical systems. Acta Mech. Sin. **28**(3), 854–862 (2012)

31. Nowakowski, C.: Zur Modellierung und Reduktion elastischer Bauteile unter verteilten Lasten für die Mehrkörpersimulation (in German). Dissertation, Schriften aus dem Institut für Technische und Numerische Mechanik der Universität Stuttgart, vol. 35. Shaker Verlag, Aachen (2014)

32. Nowakowski, C., Fehr, J., Eberhard, P.: Einfluss von Schnittstellenmodellierungen bei der Reduktion elastischer Mehrkörpersysteme (in German), at -. Automatisierungstechnik **59**(8), 512–519 (2011)

33. Fehr, J., Holzwarth, P., Eberhard, P.: Interface and model reduction for efficient explicit simulations - a case study with nonlinear vehicle crash models. Math. Comput. Model. Dyn. Syst. **22**(4), 380–396 (2016)

34. Ihrle, S., Lauxmann, M., Eiber, A., Eberhard, P.: Nonlinear modelling of the middle ear as an elastic multibody system-applying model order reduction to acousto-structural coupled systems. J. Comput. Appl. Math. **246**, 18–26 (2012)

35. Ihrle, S., Eiber, A., Eberhard, P.: Modeling of the incudo-malleolar joint within a biomechanical model of the human ear. Multibody Syst. Dyn. **39**(4), 1–20 (2016)

36. Simpack, A.G.: Multibody simulation software (2017). http://www.simpack.com/. Accessed 08 May 2017

37. FIRST, Simulationstool für elasto-hydrodynamisch gekoppelte Mehrkörpersysteme (in German). IST Ingenieurgesellschaft für Strukturanalyse und Tribologie mb. http://www.ist-aachen.de/first. Accessed 08 May 2017

38. Nowakowski, C., Fehr, J., Eberhard, P.: Model reduction for a crankshaft used in coupled simulations of engines. In: Proceedings of the ECCOMAS Thematic Conference on Multibody Dynamics 2011, Brussels, Belgium (2011)

39. Shiiba, T., Fehr, J., Eberhard, P.: Flexible multibody simulation of automotive systems with non-modal model reduction techniques. Veh. Syst. Dyn. **50**(12), 1905–1922 (2012)

40. Grunert, D., Fehr, J.: Identification of nonlinear behavior with clustering techniques in car crash simulations for better model reduction. Adv. Model. Simul. Eng. Sci. **3**(1), 1–19 (2016)

41. Bohn, B., Garcke, J., Iza-Teran, R., Paprotny, A., Peherstorfer, B., Schepsmeier, U., Thole, C.A.: Analysis of car crash simulation data with nonlinear machine learning methods. Procedia Comput. Sci. **18**, 621–630 (2013)

42. Holzwarth, P., Eberhard, P.: Input-output based model reduction for interconnected systems. Eur. J. Mech. A/Solids **49**, 408–418 (2015)

Model Order Reduction a Key Technology for Digital Twins

Dirk Hartmann, Matthias Herz and Utz Wever

Abstract An increasing number of disruptive innovations with high economic and social impact shape our digitalizing world. Speed and extending scope of these developments are limited by available tools and paradigms to master exploding complexities. Simulation technologies are key enablers of digitalization. They enable digital twins mirroring products and systems into the digital world. Digital twins require a paradigm shift. Instead of expert centric tools, engineering and operation require autonomous assist systems continuously interacting with its physical and digital environment through background simulations. Model order reduction (MOR) is a key technology to transfer highly detailed and complex simulation models to other domains and life cycle phases. Reducing the degree of freedom, i.e., increasing the speed of model execution while maintaining required accuracies and predictability, opens up new applications. Within this contribution, we address the advantages of model order reduction for model-based system engineering and real-time thermal control of electric motors.

Keywords Model order reduction · Virtual sensor · Systems engineering Krylov methods · Response surfaces

1 Introduction

Today, we are at a tipping point in engineering and operation of products and complex systems. Growing complexity (e.g., higher degree of integration in complex system, heterogeneous macro-systems or novel manufacturing technologies) and increased business pressures (e.g., accelerating innovation pace) result in higher demand and new challenges for model-based software tools in engineering and operation [25]. At the same time, available expertise (e.g., trained experts) is limiting the expansion of model-based approaches [4]. For this reason, novel paradigms for computer-aided engineering and operation are required.

D. Hartmann (✉) · M. Herz · U. Wever
Siemens AG, Corporate Technology, 80200 Munich, Germany
e-mail: hartmann.dirk@siemens.com

© Springer International Publishing AG, part of Springer Nature 2018
W. Keiper et al. (eds.), *Reduced-Order Modeling (ROM) for Simulation and Optimization*,
https://doi.org/10.1007/978-3-319-75319-5_8

167

During the last two years, the digital twin vision has emerged as a novel paradigm fruitioning technical advancements. It is one of the top ten future strategic technology trends for digitalization [18]. The digital twin is the collection of all actionable system-related information and data of a system that is updated and upgraded through all stages of its life cycle from its ideation to its end of operations [7]. For products equipped with embedded functionality, assistance for day-to-day decisions can be provided to increase reliability, quality, and efficiency of systems [18].

Model order reduction is one core technology for the digital twin vision since it allows compressing simulation models for real-time simulation as well as model exchange without intellectual property risks. Ultimately, it enables reusing simulation models from the early phases of product development in later product lifetime phases, especially during the product operation phase.

In this article, we address some particular applications where model order reduction techniques can contribute toward the digital twin vision. Furthermore, we outline some concrete industrial examples where model order reduction was successfully used within this context.

2   Digital Twin Vision

Classically simulation is applied for design and engineering issues in early phases of product and production development to support design planning and dimensioning. However, within the emerging digital twin vision simulation for the whole life cycle is needed. Simulation will not only support engineers in early development phases or later test phases but also support operators and users in parallel to the operation phase, e.g., by model-based condition monitoring and optimized control culminating in new simulation-based maintenance services. This requires novel dynamic software models to be integrated in existing simulation workflows which allow to store and access the present expert engineering knowledge in an executable manner. It is exactly this mapping and storage of existing expert knowledge which makes the digital twin vision valuable. Knowledge is reusable by non-experts and in other stages in the product life cycle; see Fig. 1.

In the rest of this section, we concentrate on two specific examples to make the concept of the digital twin concrete, namely the exchange of models between different departments during design and engineering enabling more efficient workflows as well as the transfer of engineering models to the operation phase enabling more efficient operations.

Model Exchange from 3D Simulation to System Simulation

Today simulation is an established technology, in nearly all engineering departments. However, very often simulation is used only in silos. A key challenge of simulation is

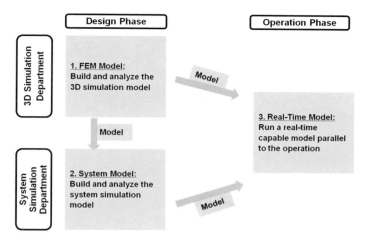

Fig. 1 Model transfer (green arrows) in the digital twin vision

the reuse of simulation in other domains [4]. For example, it is a major road blocker for a wider adoption of model-based systems engineering (MBSE) [14, 19]. MBSE uses digital system models capturing the interconnected behavior of all sub-systems on a system level. The virtual discipline-specific models and system models are the basis for a validation of the required product behavior. Using these incremental and continuous adapted models, virtual test scenarios derived from requirements are simulated during the whole research and development (R&D) process. Thus, an early correlation between system requirements, functions, and behavior is possible.

Model-based systems engineering distinguishes between different levels of simulation [6]:

- **System simulation**: Models, often very simple, simulating primary behavior (e.g., MATLAB, Modelica) and covering multiple disciplines including software and control are the basis for a systematic investigation of the developed solutions on a system level.
- **Discipline-specific model building and simulation**: Discipline-specific 3D geometry or 3D computer-aided engineering (CAE) models (e.g., Siemens Simcenter, Comsol, or ANSYS multi-physics) of high complexity and accuracy are used for validation and optimization of detailed aspects.

However, in today's simulation practice the 3D simulation department and the system simulation department are separated in the sense that each department builds the simulation models internally with commercial or in-house software tools that are available in the respective domain. However, once a detailed 3D simulation model is build up and calibrated it would be desirable to compress the contained knowledge and to reuse it as custom model for a component in a system simulation model. Industrial practice today is still manually rebuilding and updating the system model manually. This is exactly where model order reduction comes into play since it allows

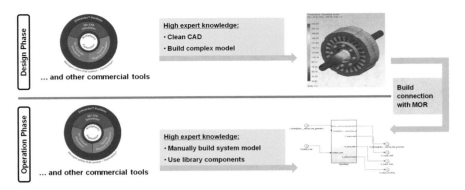

Fig. 2 Model exchange from 3D simulation to system simulation

realizing the compression step. Together with the vendor-neutral functional mock-up interface (FMI) standard [3] (as of today implemented in more than 100 tools), model order reduction algorithms can take the detailed 3D simulation model as input and provide a compressed model in a standardized format for system simulation software as output; see Fig. 2. The key point here is that model order reduction not only enables the model transfer but also allows automating the model transfer.

Model Exchange from Design Phase to Operation Phase

With growing available computing power as well as the evolution of mathematical algorithms, computer-aided paradigms are also being established as a key technology for operation. Comparing the ratio of potential benefits versus development and installation costs computer-aided approaches are far superior to classical approaches. For example, improving trucks energy efficiency by means of advanced control solutions has a significantly better improvement to cost ratio than aerodynamic measures [10].

On the one hand, machine learning approaches are replacing today's concepts based on expertise and heuristics by learning from increasing amount of sensor data. However, in an industrial context machine learning is often limited by the availability of data, e.g., many customers data carries intellectual property information inherently or addressing specific failures not enough data could be collected in the past. On the other hand, model predictive approaches can improve operation significantly, e.g., [10]. However, they require high efforts from experts triggered by the fact that simulation models which are used parallel to operation must be faster than real-time. Simulation models which are used during the product design are typically computationally expensive and need a significant solution time. Thus corresponding models need to be realized additionally and corresponding approaches are restricted to high value or high lot size applications to ensure a business case.

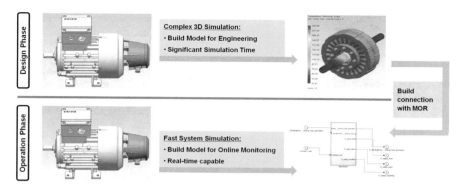

Fig. 3 Model exchange from product design to product operation

Model order reduction algorithms can compress even huge 3D simulation models in such a way that the resulting reduced-order models are small enough for faster than real-time evaluation; see Fig. 3. Thereby model order reduction allows coping with limited data and avoids the need to regenerate information in the operation phase with data-based methods. This enables for example providing new services for predicting faults, increasing operational efficiency, or for service planning [18].

3   Model Order Reduction

At the heart of the digital twin vision is continuous background simulation enabling novel assist systems during all life cycle phases. As outlined above, model order reduction is a key technology to achieve real-time capabilities. Within this contribution, we focus on model order reduction for (linear) 3D partial differential equations, which are widely used during detailed engineering. After spatial discretization, huge linear sparse equations are obtained, which often exceed 10^6 unknowns (Fig. 4). By splitting computations in an offline and an online phase, computational effort is shifted to an offline phase allowing interactive and real-time capable simulation during the online phase. A variety of concepts and approaches have been introduced in the last decades mostly using projection-based approaches such as proper orthogonal decomposition (e.g., [15, 24]), balanced truncation (e.g., [8]), reduced basis method (e.g., [20, 21]), or Krylov subspace methods (e.g., [1]). The key idea of most approaches is to reduce the space of considered functions by means of an appropriate low-dimensional basis. For (close to) linear models, model order reduction is state of the art in computational engineering and science. However, for nonlinear models it is a highly active field of research.

Using model order reduction as technology for compressing simulation models requires that for a given model the equations describing this model as well as the solution vector are compressed. For example, if a model is initially given as

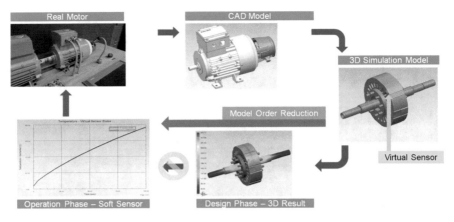

Fig. 4 Model order reduction (green) in the product life cycle simulation chain enables to go beyond today's approaches (blue)

$$F(x) = b \qquad \text{where } F : \mathbb{R}^n \to \mathbb{R}^n \text{ and } x, b \in \mathbb{R}^n \tag{1}$$

then the reduced-order model must be given as

$$F_r(x_r) = b_r \qquad \text{where } F_r : \mathbb{R}^m \to \mathbb{R}^m \text{ and } x_r, b_r \in \mathbb{R}^m. \tag{2}$$

Here the dimension m of the reduced model should be significantly smaller than the dimension n of the initial model. This simple examples demonstrates that model order reduction must compress the initial model equations described by $F(\cdot)$ to a reduced set of equations given by $F_r(\cdot)$. For this reduced set of equations, the orginal state x is approximated by the reduced state x_r. The connection between x and x_r is often given by a projection matrix $Q_r \in \mathbb{R}^{m \times n}$, i.e., $x_r = Q_r x$. Furthermore, having the compressed equations given by $F_r(\cdot)$ ensures that the compressed model can be repeatedly evaluated even for changing conditions given by b_r. For model exchange and reusing models, the crucial requirement is compressing $F(\cdot)$ to $F_r(\cdot)$ since the model equations are exactly represented by these functions.

This requirement sounds trivial but the field of model order reduction comprises of several algorithm classes which were developed with different areas of application in mind. For some of these areas, it was more the goal of compressing the orginal state x to a reduced state x_r. Hence for the purpose of exchanging compressed models, each of the existing algorithm classes must be reviewed how much they target toward compressing the model equations $F(\cdot)$ to $F_r(\cdot)$ or if they are focusing on compressing the state x to x_r.

Krylov Subspace Methods

Gathering these requirements leads to the conclusion that model order reduction algorithms based on Krylov subspaces seem to be a good starting point toward the digital twin vision. On the one hand, linear equation solvers based on Krylov subspaces are heavily used today and are well established for solving 3D simulation models based on partial differential equations. On the other hand, Krylov subspace methods aim at reducing the given model equations to a significantly smaller set of model equations from which the reduced state can be easily computed even when the system inputs $\mathbf{u}$ change.

Starting with a given state-space systems of the form

$$\mathbf{E}\dot{\mathbf{x}} = \mathbf{A}\mathbf{x} + \mathbf{B}\mathbf{u}, \tag{3}$$

$$\mathbf{y} = \mathbf{Z}\mathbf{x} + \mathbf{D}\mathbf{u}, \tag{4}$$

where $\mathbf{x} \in \mathbb{R}^n$, $\mathbf{u} \in \mathbb{R}^p$, $\mathbf{y} \in \mathbb{R}^q$, and $\mathbf{E}, \mathbf{A} \in \mathbb{R}^{n \times n}$, $\mathbf{B} \in \mathbb{R}^{n \times p}$, $\mathbf{Z} \in \mathbb{R}^{q \times n}$, $\mathbf{D} \in \mathbb{R}^{q \times p}$. Krylov methods aim to compute projection matrices $\mathbf{V}, \mathbf{W} \in \mathbb{R}^{n \times m}$ which, applied to the initial system, lead to a reduced state-space system given by

$$\mathbf{E}_r\dot{\mathbf{x}}_r = \mathbf{A}_r\mathbf{x}_r + \mathbf{B}_r\mathbf{u}, \tag{5}$$

$$\mathbf{y} = \mathbf{Z}_r\mathbf{x}_r + \mathbf{D}\mathbf{u}, \tag{6}$$

where $\mathbf{x}_r \in \mathbb{R}^m$, $\mathbf{u} \in \mathbb{R}^p$, $\mathbf{y} \in \mathbb{R}^q$ and $\mathbf{E}_r, \mathbf{A}_r \in \mathbb{R}^{m \times m}$, $\mathbf{B}_r \in \mathbb{R}^{m \times p}$, $\mathbf{Z}_r \in \mathbb{R}^{q \times m}$. As Eqs. (3)–(4) also Eqs. (5)–(6) are a state-space model, they have a standardized format and thus can be inserted into various system simulation tools (e.g., MATLAB or Amesim). For example, in the past we have successfully applied the Krylov subspace method to compress 3D simulation models with dimensions of, e.g., $n = 3.5 \cdot 10^6$ to a reduced-order model of dimension $m = 20$. Executing the reduction just once, we could evaluate the reduced model for several inputs $\mathbf{u}$. Hence, the reduced-order models based on Krylov subspace methods are able to produce faster than real-time responses for chancing system inputs $\mathbf{u}$, i.e., enabling simulation parallel to operation.

Further Methods

Krylov subspace methods are a very general method which works, e.g., for heat transfer and structural mechanics. Nevertheless it should be mentioned that modal reduction [2] is one of the first methods which was successfully applied to linear structural mechanics problems. It is included in commercial tools such as NX-Nastran. Furthermore, the outlined Krylov subspace methods work only if the underlying partial differential equation is linear. For nonlinear equations, the development of suitable methods is still under active research. One possibility to cope with nonlinear partial differential equations is to generate standardized quadratic of even cubic state-space models; see, e.g., [5, 23].

$$\mathbf{E}_r \dot{\mathbf{x}}_r = \mathbf{A}_r \mathbf{x}_r + (\mathbf{C}_r : \mathbf{x}_r) : \mathbf{x}_r + \mathbf{B}_r \mathbf{u} \tag{7}$$

where $\mathbf{C}_r \in \mathbb{R}^{m \times m \times m}$. Current methods for reducing nonlinear equations work in another way. They assume the knowledge of the design space and use discrete samples in order to set up a surrogate function, e.g., via neural networks [9] or simple interpolation [11]. Another way to make use of the discrete samples or snapshots is the application of proper orthogonal decomposition [17]. Here, a suitable base is extracted in terms of the leading eigenvectors of an eigenvalue problem. Either the projection could be used systematically to set up the reduced model [17] or the reduced model is simply learned by machine learning [16]. Furthermore, a posteriori error estimation is a very active of research for reduced-order modeling. Thus, not only can the models be reduced but also detailed error estimates are available for linear as well as nonlinear problems and control problems.

4 Applications

Use Case System Optimization—Model-Based Systems Engineering

To highlight the advantages of standardized model order reduction solutions in a systems engineering context, we consider an electronic module with controllable air cooling. The electronic module (Fig. 5a) is heated up due to the electric power. Some parts of the electronic module are sensible to high temperatures, and they cannot sustain a temperature (e.g., 67 °C) over a longer time. Two temperature sensors are installed in the electronic module to monitor thermal behavior. To prevent exceeding the critical temperature over a longer time, e.g., due to external temperature fluctuations or varying electric powers, a control for an efficient operation (energy saving and quiet) of the electronic module shall be realized.

On the basis of a CFD simulation (Fig. 5b), the thermal behavior of the module for given airflow can be predicted. Such simulation models are typically set up for a validation of the detailed design. Principally, these simulations could be also used for realizing control strategies. However, due to the required computational efforts for the corresponding 3D simulations, this is typically not realized and simplified low-fidelity models are build up manually to verify control strategies. Apart from the manual effort to realize corresponding solutions, there is also a risk of manually built low-fidelity models not being updated according to design changes.

Here, we take an approach based on model order reduction. For simplicity, we have used response surface models. These can be easily implemented and also nonlinear

(a)

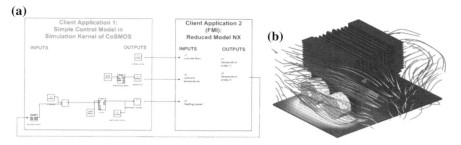

(b)

Fig. 5 a Development and validation of a control on the basis of a system simulation using the proprietary simulation platform CoSMOS [22] as the co-simulation masters. **b** The reduced model is derived by means of model order reductions from a CFD simulation in NX 8.5 and encapsulated as an FMI model. Goal is an optimal control of the air cooler speed to allow for an efficient cooling (energy saving and loudness as well as ensuring that the temperature at critical point is not exceeded for a longer time)

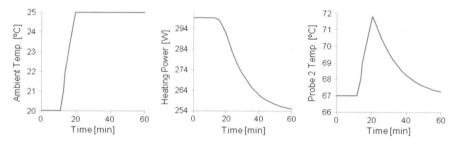

Fig. 6 Results of the system simulation: An increase in the environmental temperature leads to a reduction of the electric power. The critical temperature of 67 C is not exceeded over a longer time [11]

effects can be captured. The models could be setup with little effort in other disciplines or phases (here: development of controls). For more details, we refer to [11].

Based on the reduced model, a functional mock-up unit (FMU) is generated according to the functional mock-up interfaces (FMI) standard [3]. Being supported by more than 100 commercial and open source tools, the generated FMU can be integrated in most system simulators. Here, the proprietary system simulation platform CoSMOS [22] is used as the basis for development and validation of a control strategy (c.f. Fig. 5). For simplicity, we have set up only a control using one temperature as an input and the controlled electric power as an output. The result is shown in Fig. 6.

Use Case Simulation Augmented Assistance Thermal Control of an Electric Motor

As a second use case, the efficient thermal control of an asynchronous electric motor is considered. Electrical motors are subject to thermal derating requiring controller-managed temperatures [12]. For example, large-scale asynchronous electric drives are exposed to high levels of stress due to induction heating on start-up. Frequent starts without a sufficient interval for cooling can result in motors overheating. However, measuring the temperature of the actual rotors turning at high speed within the motor is close to impossible. Thus, controls are often based on very conservative heuristics.

In principle, corresponding temperatures can be calculated by means of 3D thermal simulations sufficiently accurate. In the case of the electric motor considered here linear convection–diffusion models are used. Corresponding 3D models are typically available from detailed engineering. However, they are computationally too demanding to be used during operation. Using model order reduction corresponding models for background simulation can be realized. Here, we are relying on Krylov methods as introduced before. Being highly efficient to evaluate they allow not only to be extended by uncertainty quantification methods but also support continuous calibration using available sensors on the stator side. Thus, by means of continuously calibrated background simulation models, temperatures not accessible to sensors can be virtually measured. Using, e.g., augmented reality devices the user can at anytime virtually look inside the motor to observe temperature distributions; c.f. Fig. 7a [13]. This allows the cooling times required for electric motors to be significantly reduced, ultimately enhancing plant availability. Enriched with methods for uncertainty quantification, confidence intervals can be provided for the rotor temperature (c.f. Fig. 7b) allowing to go close to operational limits.

To underline the accuracy of the reduced-order model, we have computed the temperature curve at the sensor position based on the reduced-order model as well

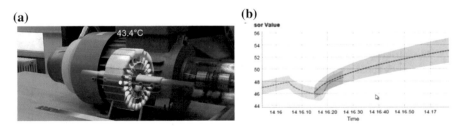

(a) **(b)**

Fig. 7 **a** Temperature visualization inside an electric motor by means of augmented reality using Microsoft Hololens [13]. **b** Screenshot of the virtual temperature sensor solution estimating current rotor temperature (green) as well as predicting rotor temperature after start-up (blue) at a specific point. The plotted bands show the calculated uncertainties [12]

Fig. 8 Temperature curve based on the full 3D model (red) and the reduced-order model (blue)

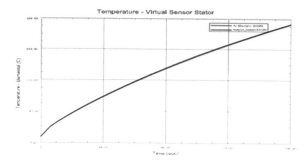

as based on the full 3D model. Overlaying these predictions shows that model order reduction is able to keep the predicted behavior in the reduction process up to the required accuracies; see Fig. 8.

5 Conclusions

Model order reduction is probably the most important mathematical technique for realizing a digital twin during the life cycle of a product. It enables new levels of interactivity, reliability, continuity, accessibility, and distributability of simulation models:

- **Interactivity**: There are many methods how to speed up 3D simulations. Among the most successful approaches are, e.g., the exploitation of computer architectures or advanced solver technologies. Model order reduction directly reduces the number of degrees of freedom and therefore is able to generate arbitrary fast models. For real-time applications, a suitable compromise between accuracy and speed can be realized, e.g., by adaptive methods.
- **Reliability**: In most cases, the speedup of simulations tools lead to a lack of accuracy. However, this lack of accuracy should be quantified, because the user wants to know the expected accuracy of the reduced model. For many model order reduction schemes, detailed error analyses are known, e.g., [2, 20]. Thus, error bounds and confidence intervals (when calibrated by data) can be provided using reduced models.
- **Continuity**: Seamless workflow of simulation tools during the life cycle is still a major vision since many years. For the design, a lot of manual effort has to be spent in order to obtain accurate models. On system-level simulation and for real-time applications, very fast models are required for exactly the same component, again requiring high manual efforts today. The technique of model order reduction exactly bridges the different applications, i.e., model order reduction aims to automatically generate fast models from comprehensive design models.

- **Accessibility**: Setting up complex models for design is still a task for dedicated experts. However, the generated fast models via Model Order Reduction can be mashed behind standardized formats and can thus be integrated into system simulation also by non-experts. This point is a central aspect in the digital twin vision.
- **Distributability**: Solving the fast models generated by model order reduction requires only matrix–vector multiplications. Thus, they may run also on very small computers (and not only on specialized workstations). Furthermore, they do not allow reconstructing the component from the model as it is possible from the computer-aided design (CAD) or meshed computer-aided engineering (CAE) models. That is, they protect intellectual property, which is another major issue. In a few years, each complex component might be delivered with a corresponding digital twin.

References

1. Bai, Z.: Krylov subspace techniques for reduced-order modeling of large-scale dynamical systems. Appl. Numerical Math. **43**(1–2), 9–44 (2002)
2. Baur, U., Benner, P., Feng, L.: Model order reduction for linear and nonlinear systems: a system-theoretic perspective. Arch. Comput. Methods Eng. **21**(4), 331–358 (2014)
3. Blochwitz, T., Otter, M., Arnold, M., Bausch, C., Elmqvist, H., Junghanns, A., Mauß, J., Monteiro, M., Neidhold, T., Neumerkel, D., et al.: The functional mockup interface for tool independent exchange of simulation models. In: Proceedings of the 8th International Modelica Conference; March 20th-22nd; Technical Univeristy; Dresden; Germany, number 063, pp. 105–114. Linköping University Electronic Press (2011)
4. CIMdata. Model-based systems engineering business opportunities and overcoming implementation challenges. Cimdata report (2014)
5. Desjardins, A., Rixen, D.J., Rutzmoser, J.B., Wever, U.: A hyper reduction technique for real time structural mechanical applications (in preparation) (2018)
6. Eigner, M., Dickopf, T., Apostolov, H., Schaefer, P., Faißt, K.G., Keßler, A.: System lifecycle management: initial approach for a sustainable product development process based on methods of model based systems engineering. In: PLM, pp. 287–300 (2014)
7. Grieves, M.: Digital twin: manufacturing excellence through virtual factory replication. White paper (2014)
8. Gugercin, S., Antoulas, A.C.: A survey of model reduction by balanced truncation and some new results. Int. J. Control **77**(8), 748–766 (2004)
9. Guo, X., Li, W., Iorio, F.: Convolutional neural networks for steady flow approximation. In: Proceedings of the 22nd ACM SIGKDD International Conference on Knowledge Discovery and Data Mining, pp. 481–490. ACM (2016)
10. Haas, B.: Predictive control systems in heavy-duty commercial vehicles. In: Proceedings of Automotive Powertrain Control Systems (2012)
11. Hartmann, D., Mahler, M.: Automatische Generierung von standardisierten Systemmodellen aus 3d-simulationen im systems Engineering Kontext. NAFEMS Online Magazin, 33 (2014)
12. Hartmann, D., Obst, B.: Taking the heat off - Simulation to reliably determine motor temperature. Siemens Industry News (2016)
13. Hartmann, D., Papadopoulos, T.: Virtual X-ray for large motors. Siemens YouTube channel - https://youtu.be/86vkjykbHRM (2018)
14. Haskins, C., Walden, F.D., Hamelin, D.: Systems engineering handbook. In: Krueger, M. (ed.) INCOSE (2006)

15. Holmes, P., Lumley, J.L., Berkooz, G.: Turbulence, Coherent Structures, Dynamical Systems and Symmetry. Cambridge University Press (1996)
16. Krischer, K., Rico-Martínez, R., Kevrekidis, I.G., Rotermund, H.H., Ertl, G., Hudson, J.L.: Model identification of a spatiotemporally varying catalytic reaction. AIChE J. **39**(1), 89–98 (1993)
17. Meyer, M., Matthies, H.G.: Efficient model reduction in non-linear dynamics using the karhunen-loeve expansion and dual-weighted-residual methods. Comput. Mech. **31**(1), 179–191 (2003)
18. Panetta, K.: Gartner Top 10 Strategic Technology Trends for 2018. Gartner (2018)
19. Paredis, C.: Model-based systems engineering: A roadmap for academic research. Frontiers in Model-Based Systems Engineering, Atlanta, GA (2011)
20. Patera, A.T., Rozza, G.: Reduced basis approximation and a posteriori error estimation for parameterized partial differential equations (2007)
21. Quarteroni, A., Rozza, G., Manzoni, A.: Certified reduced basis approximation for parametrized partial differential equations and applications. J. Math. Industry **1**(1), 3 (2011)
22. Schenk, T., Gilg, A.B., Mühlbauer, M., Rosen, R., Wehrstedt, J.: Architecture for modeling and simulation of technical systems along their lifecycle. Comput. Vis. Sci. **17**(4), 167–183 (2015)
23. Weeger, O., Wever, U., Simeon, B.: On the use of modal derivatives for nonlinear model order reduction. Int. J. Numer. Meth. Eng. **108**(13), 1579–1602 (2016)
24. Willcox, K., Peraire, J.: Balanced model reduction via the proper orthogonal decomposition. AIAA J. **40**(11), 2323–2330 (2002)
25. Woo, T.: The democratization of simulation in a multiphysics world (2016)

Printed in the United States
By Bookmasters